Simulation Modeling Using @RISK

Books of Related Interest

ADA Decision Systems, *DPL Decision Analysis Software,* Student Version
Berk/Carey, *Data Analysis with Microsoft Excel*
Bowerman/O'Connell, *Forecasting and Time Series: An Applied Approach,* 3rd ed.
Bowerman/O'Connell, *Linear Statistical Models: An Applied Approach,* 2nd ed.
Canavos/Miller, *Modern Business Statistics*
Chisman, *Industrial Cases in Simulation Modeling*
Clauss, *Applied Management Science and Spreadsheet Modeling*
Clemen, *Making Hard Decisions: An Introduction to Decision Analysis,* 2nd ed.
Colley, *Case Studies in Service Operations*
Cryer/Miller, *Statistics for Business: Data Analysis and Modeling,* 2nd ed.
Davis, *Business Research for Decision Making,* 4th ed.
Devore, *Probability and Statistics for Engineering and the Sciences,* 4th ed.
Dielman, *Applied Regression Analysis for Business and Economics,* 2nd ed.
Durrett, *The Essentials of Probability*
Durrett, *Probability: Theory and Examples*
Etienne-Hamilton, *Operations Strategy for Competitive Advantage*
Farnum, *Modern Statistical Quality Control and Improvement*
Finch/Luebbe, *Operations Management: Competing in a Changing Environment*
Gaither, *Production and Operations Management,* 7th ed.
Glaskowsky/Hudson/Ivie, *Business Logistics,* 3rd ed.
Graybill/Iyer, *Regression Analysis: Concepts and Applications*
Hamburg/Young, *Statistical Analysis for Decision Making,* 6th ed.
Hamilton, *Regression with Graphics: A Second Course in Applied Statistics*
Hamilton, *Statistics with STATA,* Version 3.0
Hedderson/Fisher, *SPSS Made Simple,* 2nd ed.
Higgins/Keller-McNulty, *Concepts in Probability and Stochastic Modeling*
Hildebrand/Ott, *Basic Statistical Ideas for Managers*
Hildebrand/Ott, *Statistical Thinking for Managers,* 3rd ed.
Keller/Warrack/Bartel, *Statistics for Management and Economics,* 3rd ed.
Kenkel, *Introductory Statistics for Management and Economics,* 4th ed.
Kleinbaum et al., *Applied Regression Analysis and Other Multivariable Methods,* 2nd ed.
Kuehl, *Statistical Principles of Research Design and Analysis*
Lapin, *Cases in Management Science*
Lapin, *Quantitative Methods for Business Decisions with Cases,* 6th ed.
Lapin, *Statistics for Modern Business Decisions,* 6th ed.
Lunneborg, *Modeling Experimental and Observational Data*
Mendenhall/Beaver/Beaver, *A Course in Business Statistics,* 4th ed.
Mendenhall/Reinmuth/Beaver, *Statistics for Management and Economics,* 7th ed.
Mendenhall/Wackerly/Scheaffer, *Mathematical Statistics with Applications,* 5th ed.
Menzefricke, *Statistics for Managers*
Middleton, *Data Analysis Using Microsoft Excel 5.0*
Miller/Wichern, *Intermediate Business Statistics*
Myers, *Classical and Modern Regression with Applications,* 2nd ed.
Ott, *An Introduction to Statistical Methods and Data Analysis,* 4th ed.
Rice, *Mathematical Statistics and Data Analysis,* 2nd ed.
Roberts, *Data Analysis for Managers,* 2nd ed.
Ryan/Joiner, *Minitab Handbook,* 3rd ed.
SAS® Institute, Inc., *JMP IN 3 for Windows*
Scheaffer, *Introduction to Probability and Its Applications,* 2nd ed.
Scheaffer/McClave, *Probability and Statistics for Engineers,* 4th ed.
Scheaffer/Mendenhall/Ott, *Elementary Survey Sampling,* 5th ed.
Schrage, *LINDO: An Optimization Modeling System,* 4th ed.
Selvin, *Practical Biostatistical Methods*
Shipley/Ruthstrom, *Cases in Operations Management*
Sower/Motwani/Savoie, *Classic Readings in Operations Management*
Tanur et al., *Statistics: A Guide to the Unknown,* 3rd ed.
Van Matre, *Foundations of TQM: A Readings Book*
Weiers, *Introduction to Business Statistics,* 2nd ed.
Winston, *Introduction to Mathematical Programming,* 2nd ed.
Winston, *Operations Research: Applications and Algorithms,* 3rd ed.

For more information about any of these titles, contact your local bookseller or write: Duxbury Press, Wadsworth Publishing Company, 10 Davis Drive, Belmont, CA 94002.

Simulation Modeling Using @RISK

Wayne L. Winston
School of Business
Indiana University

Duxbury Press
An Imprint of Wadsworth Publishing Company
 An International Thomson Publishing Company

Belmont • Albany • Bonn • Boston • Cincinnati • Detroit • London • Madrid • Melbourne
Mexico City • New York • Paris • San Francisco • Singapore • Tokyo • Toronto • Washington

Editor: Curt Hinrichs
Editorial Assistant: Cynthia Mazow
Marketing Manager: Joanne Terhaar
Advertising Project Manager: Joseph Jodar
Production Editor: Sandra Craig
Print Buyer: Karen Hunt
Permissions Editor: Peggy Meehan
Copy Editor: Erin Milnes
Text Design and Composition: Penna Design & Production
Cover Design: William Reuter Design

Cover Photographs: Financial still life: U.S. dollar and stock quotations, © Frank Saragnese, FPG International; workers in automobile production plant, © Telegraph Colour Library, FPG International; composite of business symbols, Ed Honowitz, ©Tony Stone Images; roulette wheel, Ken Whitmore, ©Tony Stone Images.

Printer: Courier Companies, Inc./Stoughton

Printed in the United States of America
4 5 6 7 8 9 10

For more information, contact Duxbury Press at Wadsworth Publishing Company:

Wadsworth Publishing Company
10 Davis Drive
Belmont, California 94002, USA

International Thomson Publishing Europe
Berkshire House 168-173
High Holborn
London, WC1V 7AA, England

Thomas Nelson Australia
102 Dodds Street
South Melbourne 3205
Victoria, Australia

Nelson Canada
1120 Birchmount Road
Scarborough, Ontario
Canada M1K 5G4

International Thomson Editores
Campos Eliseos 385, Piso 7
Col. Polanco
11560 México D.F. México

International Thomson Publishing GmbH
Königswinterer Strasse 418
53227 Bonn, Germany

International Thomson Publishing Asia
221 Henderson Road
#05-10 Henderson Building
Singapore 0315

International Thomson Publishing Japan
Hirakawacho Kyowa Building, 3F
2-2-1 Hirakawacho
Chiyoda-ku, Tokyo 102, Japan

Library of Congress Cataloging-in-Publication Data

Winston, Wayne L.
Simulation modeling using @Risk/Wayne L. Winston.
p. cm.
Includes index.
ISBN 0-534-26491-3 (book with full-functioning academic software)
ISBN 0-534-26492-1 (book with demonstration version software)
1. At Risk (Computer file) 2. Business—Computer simulation.
I. Title
HF5548.2.W477 1996
658.15'5'02855369—dc20 95-40571

This book is based on materials developed for *Management Science: Applications and Spreadsheet Modeling* (ISBN 0-534-21774-5), by Wayne L. Winston and S. Christian Albright, to be published by Duxbury Press in summer 1996.

CONTENTS

PREFACE

Many excellent simulation texts have been written during the last twenty years. Most of them concentrate on teaching students how to simulate queuing and inventory systems or other systems were bottlenecks occur. These texts also require students to write simulation programs using a language such as C or Fortran or a canned simulation program such as SLAM, SIMAAN, or GPSS. Certainly "bottleneck" simulations are of great importance, but the typical MBA will probably go into a finance or marketing job where bottleneck simulations are of little interest.

Many Fortune 500 firms are adopting simulation as the method for doing capital budgeting and analyzing the introduction of new products. Wall Street uses simulation daily to price complex, "exotic" derivatives. The nice thing is that most finance and marketing simulations can easily be performed in a spreadsheet using the spreadsheet add-in @RISK.

Purpose

Simulation Modeling Using @RISK introduces MBAs and advanced undergraduates to spreadsheet business simulation models. Each chapter contains a virtually self-contained discussion of an interesting business model that can be simulated with @RISK. We run the gamut from determining the price of financial derivatives to modeling the evolution of a company's brand share over time. We also give a brief treatment of spreadsheet simulation of inventory and queuing models. Each section contains step-by-step instructions on how to perform the simulation. The examples are drawn from students' course work in finance, marketing, and production.

The book has been class tested for three years in a highly successful elective management science course at Indiana University. Students enjoy the material, find it useful for enhancing their spreadsheet and modeling skills, and a description of the course looks good on their resumes.

Two versions of the text are available: either accompanied by the full-functioning @RISK program or (at a lower cost) with a demonstration program only. Both versions include all the book's models. Over 100 problems are included. The version with the full-functioning @RISK software allows users to operate other problems. A disk containing solutions to the problems is available to adopters.

Potential Uses

The text is suitable for several types of situations:

1. As a supplement to a "bottleneck"oriented simulation text in a typical one-semester simulation course.
2. Many MBA (and some undergraduate programs) are going to the eight-week class as their primary means of instruction. This book is ideally suited as a self-contained text for an eight-week course in business application of simulation.
3. There are many exciting applications of simulation to finance. Corporate finance or investment instructors who wish to use simulation in their classes should find this book suitable as a course supplement.
4. Executives who want to understand spreadsheet simulation models should find the book an ideal vehicle for mastering spreadsheet simulation through self-study.

The only chapters that must be covered are 1–8, 13, 14, 16, and 20. Once these chapters are covered, the remaining chapters are all self-contained and may be covered in any order. During the eight weeks I spend on simulation I cover Chapters 1–8, 10, 13, 14, and 16–22. This provides a course that emphasizes finance applications. An eight-week course emphasizing operations management applications would consist of Chapters 1–16, 20, and 22–24.

Acknowledgments

This book has been lots of fun to write. I would like to thank all my K510 students for their support of my teaching approach. The following reviewers offered valuable suggestions: Erne Houghton, The University of Sydney; Fred S. Hulme, Baylor University; Joseph McCarthy, Bryant College; Eleni Pratsini, University of Miami, Ohio; Thomas J. Schriber, University of Michigan; Carl R. Schultz, University of New Mexico; and Carlton H. Scott, University of California, Irvine. Curt Hinrichs' support of the project is greatly appreciated. Sandra Craig did her usual excellent job in producing the book. Finally, I would like to thank Sam McLafferty of the Palisade Corporation for allowing us to include the great @RISK program in the book.

Please write, phone, or e-mail me if you have any questions or ideas for chapters that should be included in future editions.

Wayne Winston (812-855-3495; e-mail WINSTON@INDIANA.EDU)
Indiana University School of Business
Bloomington, IN 47405

CHAPTER 1

What is Simulation?

In many situations an **analytic model** exists that a decision maker can use to make an optimal decision. By an analytic model we mean a mathematical equation(s) that will, for given values of certain inputs, enable the user of the model to determine the value of important outputs. For example, consider a bank in which all customers wait in a single line for the first available teller. Consider the following inputs:

1 Average number of arrivals per hour.

2 Average number of customers that a teller can serve in an hour.

3 Number of servers.

Under some reasonable conditions, some mathematical equations (see Chapter 22 of Winston 1994) can be derived. If we know Inputs 1–3, then these equations can be used to compute the following outputs:

1 Expected time a customer spends in line.

2 Expected time a customer spends in the bank.

3 Expected number of customers in line.

4 For any t, the probability that a given customer will spend more than t hours in line.

We can use this model to see how the behavior of the bank responds to a change in the number of tellers or the average time a teller requires to serve a customer. This analysis can eventually lead to a decision about the number of tellers the bank should hire.

In many situations in which uncertainty is present, however, it is difficult (or impossible) to build a tractable analytic model that will yield useful information to a decision maker. For example, consider a supermarket in which customers may freely jockey between express and regular checkout lines. In this case there are no formulas that can enable us to determine Outputs 1–4 from knowledge of Inputs 1–3. Simulation is used in situations like this where no tractable mathematical model exists.

In most instances, a **simulation model** is a computer model that imitates a real-life situation. The meaning of the word *imitates* will become clear to readers as they work through the examples of this book. Often the simulation model can provide a decision maker with important information. For our supermarket illustration, a simulation model might help us answer questions such as the following:

- If an additional express lane were added, how much would the average waiting time of a customer decrease?
- Currently baggers help people load their groceries into their car. How much would the average waiting time of a customer decrease if we eliminated this practice?
- How does the number of checkout counters needed to provide adequate service vary during the day?

This information would be a great aid in scheduling employees.

A simulation can be used to determine how sensitive a system is to changes in operating conditions. For example, if the store experiences a 20% increase in business, what will happen to the average time customers must wait for service?

A simulation model also allows the user to determine an "optimal" operating policy. For example, the supermarket could use the simulation to determine the relationship between number of open registers and expected time a customer spends waiting in line. Then this relationship (along with the cost of opening a register) could be used to determine for any customer-arrival rate the number of registers that need to be open to minimize the expected cost per unit time.

1.1 Actual Applications of Simulation

There are many published applications of simulation. Several are listed in the references at the end of the chapter. We now briefly describe several applications of simulation.

Burger King (see Swart and Donno 1981) developed a simulation model for its restaurants. This model was used to justify business decisions such as the following:

- Should a restaurant open a second drive-through window?
- How much would customer waiting time increase if a new sandwich is added to the menu?

Many companies (Cummins Engine, Merck, Procter & Gamble, Kodak, and United Airlines, to name a few) use simulation (often referred to as **risk analysis**; see Hertz 1964) to determine which of several possible investment projects should be chosen.

Consider a situation in which a company must choose a single investment. If the future cash flows for each investment project are known with certainty, then most companies advocate choosing the investment with the largest Net Present Value (NPV). If future cash flows are not known with certainty, however, then it is not clear how to choose between competing projects.

Using simulation, you can obtain a frequency distribution, or **histogram**, for the NPV of a project. You can answer questions such as these:

- Which project is riskiest?
- What is the probability that an investment will yield at least a 20% return?
- What is the probability that the investment will have an NPV of less than or equal to $1 billion?

As we will see in Chapter 6, contrary to what many finance textbooks assert, the investment with the largest expected Net Present Value may not always be the best investment!

To illustrate the use of simulation in corporate finance we refer the reader to the article "A New Tool to Help Managers" in the May 30, 1994, issue of *Fortune*. This article describes a simulation model (referred to as a Monte Carlo model) that was used by Merck (the world's largest pharmaceutical company) to determine whether Merck should pay $6.6 billion to acquire Medco, a mail-order drug company. The model contained inputs concerning the following:

- Possible futures of the U.S. health-care system such as a single-payer system, universal coverage, and so forth.
- Possible future changes in the mix of generic and brand-name drugs.
- A probability distribution of profit margins for each product.
- Assumptions about how competitors would behave after a merger with Medco.

The Merck model contained thousands of equations. A simulation was performed to see how the merger would perform under various scenarios. As Merck's CFO, Judy Lewent, said,

> Monte Carlo techniques are a very, very powerful tool to get a more intelligent look at a range of outcomes. It's almost never useful in this kind of environment to build a single bullet forecast.

Merck's model indicated that the merger with Medco would benefit Merck no matter what type of health insurance plan (if any) the federal government enacted.

1.2 What's Ahead?

In this book we will explain how to use spreadsheets to build simulation models of diverse situations. We will use the Lotus 1-2-3 and Excel add-in @RISK to perform most of our simulations. When readers have worked through the examples in the book they should have accomplished the following:

- Set up simulation models of complex real-world situations.
- Run a simulation model with @RISK.
- Interpret the simulation output and relate it to the real-world situation that is being modeled.

The book is laid out in twenty-four chapters. In Chapter 2 we define and discuss random numbers, the building blocks of simulation. In Chapter 3 we show how to use the random number generation and data table capabilities of spreadsheets to simulate a simple newsvendor problem (e.g., how many calendars should a bookstore order in the presence of uncertain demand?). This chapter also introduces some important ideas involved in the statistical analysis of simulations.

In Chapters 4 and 5 we introduce the extensive capabilities of @RISK by using @RISK to simulate several variations of the newsvendor problem.

In Chapter 6 we show how a simulation model can be used to estimate the probability distribution of the NPV earned by an investment project. We see that the project with the highest expected NPV may not always be the best project.

All firms need to set up a cash budget. In Chapter 7 we show how simulation can help a firm answer questions such as the following:

- What is the probability that the firm will lose money this year?
- What is the probability that the firm will have to borrow more than $1 million in the next year?

All firms must determine the proper production capacity level for their products. Too little capacity results in lost profits, while too much capacity causes the firm to incur excessive fixed costs. In Chapter 8 we show how past sales data can be used to build a simulation model that enables a firm to determine the proper capacity level.

Suppose you are bidding on a contract against several bidders. If you bid too low you will probably get the contract, but make very little money on the contract. If you bid too high, you will probably not get the contract. How much should you bid? Bidding problems are discussed in Chapter 9.

Suppose your goal is to shoot a cannon and hit a target one mile away. If your first cannonball falls 20 feet short of the target, most people feel that you should adjust the settings of the cannon so that the next shot will go longer. This is usually not the case. In Chapter 10 we discuss Edwards Deming's famous funnel experiment. As we shall see, the moral of the funnel experiment is "If it isn't broken, don't fix it."

Suppose one company's product meets specifications 100% of the time and another company meets specifications 95% of the time. Which company's product is of higher quality? Unless you understand the Taguchi loss function (discussed in Chapter 11), the answer may surprise you.

In Chapter 12 we show how simulation can be used to extend critical path analysis of project networks so that the following questions can be answered:

- What is the probability distribution of the time needed to complete a project?
- What is the probability that a given activity will be a critical activity?

In Chapter 13 we show how simulation can be used to determine the probability of winning for games such as craps. Also in this section, the student will learn how to use nested `IF` statements to express complex relationships.

A key decison in manufacturing is how often to replace machinery. In Chapters 14 and 15 we show how simulation can be used to analyze the effectiveness of various replacement policies. In Chapter 14 we also illustrate the use of spreadsheet `LOOKUP` tables.

Simulation is widely used in the investment industry. In Chapters 16–19 we show how to simulate the returns on portfolios involving stocks, options, futures, and bonds. For example, we can answer questions such as these:

- What is the probability that an investor's portfolio will earn an annual return exceeding 10%?
- What is the probability that an investor's portfolio will lose more than 20% during the course of a year?
- How do we determine the interest rate risk associated with a portfolio of bonds?

Derivatives (such as futures and stock options) are securities that derive their value from the price of an underlying asset. In Chapter 16 we show (see Hull 1993) how to use simulation to determine the fair market value of a derivative such as a put or call option. In Chapter 17 we show how to use @RISK to price so-called exotic options whose value depends on the price of the stock at different points in time. For example, an Asian option's value depends on the average weekly price over a 52-week period.

A company's market share is uncertain. In Chapters 20 and 21 we show how to use simulation to determine how marketing decisions such as advertising, promotions, and product characteristics influence market share and profit. In Chapter 21 we show how @RISK can be used to generate correlated random variables.

Manufacturing companies often assess the quality of a shipment from a supplier by sampling the shipment. In Chapter 22 we show how simulation can be used to determine the effectiveness of various sampling plans.

Companies facing uncertain product demand must make two important inventory decisions:

- How low should the firm let its inventory go before it reorders the product?
- What should be the size of each order?

In Chapter 23 we show how @RISK can be used to answer these questions.

How does a customer's wait in line depend on the arrival rate of customers and the service time distribution of customers? Such questions are considered in our discussion of @RISK queuing models in Chapter 24.

After you have worked through the book, you will be able to apply simulation to a wide variety of situations involving decision making under uncertainty. You should even have gained enough expertise to formulate a model of many situations not covered in the book. We urge you to attempt as many homework problems as possible. Group A problems are fairly routine, while Group B problems are more difficult. Have fun!

1.3 Simulation Models Versus Analytic Models

It must be emphasized that a simulation model provides only an approximate answer to a problem. For example, in Chapter 13 we implement a spreadsheet simulation of craps. For the 900 times we played craps, we won 47.7% of the time. This is an estimate of the probability of winning at craps. By analytic methods (using the theory of Markov chains) we can show that the actual probability of winning at craps is .493. If we ran our craps simulation many, many times, it can be shown that we would closely approximate the actual probability of winning at craps, but we could never be sure we had found the exact probability of winning at craps. For this reason, an analytic solution is preferable to a simulation solution. The problem is that for many situations an analytic solution does not exist.

We have also found that most business school students have difficulty understanding the mathematical underpinnings of analytic solutions to situations involving uncertainty. Simulation, on the other hand, is a very intuitive tool that students take to easily. Whereas only the best students can develop an analytic solution to a new situation, most students have little trouble extending the simulation approach to a new situation. In fact, many students who have studied the material in this manuscript have gone on to build useful @RISK models of *real* problems facing Fortune 500 corporations. I'm sure you will follow in their footsteps!

References

Hertz, D. 1964. "Risk Analysis in Capital Investment." *Harvard Business Review* 42 (Jan.–Feb.): 96–108.

Hull, J. 1993. *Options, Futures, and Other Derivative Securities*. Englewood Cliffs, NJ: Prentice-Hall.

Norton, R. 1994. "A New Tool to Help Managers." *Fortune* (May 30): 135–40.

Swart, W., and L. Donno. 1981. "Simulation Modeling Improves Operations, Planning, and Productivity of Fast Food Restaurants." *Interfaces* 11 (no. 6): 35–47.

Winston, W. 1994. *Operations Research: Applications and Algorithms*. Belmont, CA: Duxbury.

CHAPTER 2

Random Numbers: The Building Blocks of Simulation

Any spreadsheet package is capable of generating *random numbers* that are between 0 and 1. In the next two sections we will show how random numbers can be used to model the uncertainty inherent in many business situations. With Excel you generate a random number between 0 and 1 by typing `=RAND()` in any cell. To generate a random number with Lotus 1-2-3 simply type `@RAND` in any cell. In addition to falling between 0 and 1, the numbers created by the `=RAND()` (or `@RAND`) command have two properties that make them truly random numbers:

Property 1 Each time the `@RAND` command is used, any number between 0 and 1 has the same chance of occurring. This means, for example, that approximately 10% of the time we use `@RAND` we will get a number between 0 and .1; 10% of the time we will get a number between .65 and .75; 60% of the time we will get a number between .20 and .80, etc. Property 1 is often expressed thus: the random numbers produced by `=RAND()` or `@RAND` are *uniformly distributed* on the interval (0, 1). We call such random numbers ***U*(0, 1) random numbers**.

Property 2 (the Independence Property) Different random numbers generated by the computer are independent. This implies, for instance, that if we generate a random number in cell A5 and know its value, then this tells us nothing about the values of any other random numbers generated in the spreadsheet. Thus if `=RAND()` or `@RAND` yielded a large (say .98) random number in cell A5, then there would still be a 50% chance that `=RAND()` or `@RAND` in A6 (or any other cell) would yield a value of .50 or less.

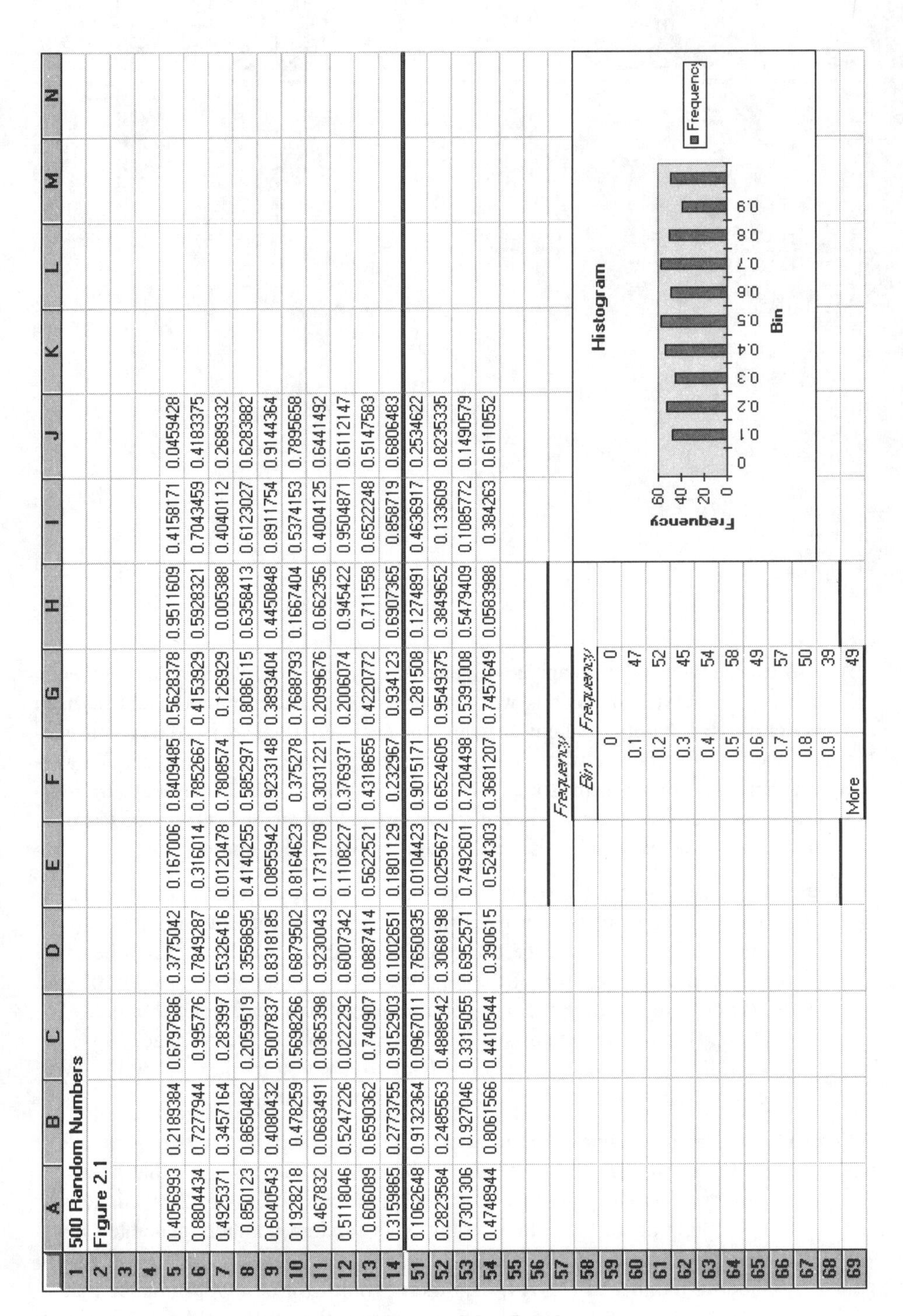

	A	B	C	D	E	F	G	H	I	J
1	500 Random Numbers									
2	Figure 2.1									
3										
4										
5	0.4056993	0.2189384	0.6797686	0.3775042	0.167006	0.8409485	0.5628378	0.9511609	0.4158171	0.0459428
6	0.8804434	0.7277944	0.995776	0.7849287	0.316014	0.7852667	0.4153929	0.5928321	0.7043459	0.4183375
7	0.4925371	0.3457164	0.283997	0.5326416	0.0120478	0.7808574	0.126929	0.005388	0.4040112	0.2689332
8	0.850123	0.8650482	0.2059519	0.3558695	0.4140255	0.5852971	0.8086115	0.6358413	0.6123027	0.6283882
9	0.6040543	0.4080432	0.5007837	0.8318185	0.0855942	0.9233148	0.3893404	0.4450848	0.8911754	0.9144364
10	0.1928218	0.478259	0.5698266	0.6879502	0.8164623	0.375278	0.7688793	0.1667404	0.5374153	0.7895658
11	0.467832	0.0683491	0.0365398	0.9230043	0.1731709	0.3031221	0.2099676	0.662356	0.4004125	0.6441492
12	0.5118046	0.5247226	0.0222292	0.6007342	0.1108227	0.3769371	0.2006074	0.945422	0.9504871	0.6112147
13	0.606089	0.6590362	0.740907	0.0887414	0.5622521	0.4318655	0.4220772	0.711558	0.6522248	0.5147583
14	0.3159865	0.2773755	0.9152903	0.1002651	0.1801129	0.232967	0.934123	0.6907365	0.858719	0.6806483
51	0.1062648	0.9132364	0.0967011	0.7650835	0.0104423	0.9015171	0.281508	0.1274891	0.4636917	0.2534622
52	0.2823584	0.2485563	0.4888542	0.3068198	0.0255672	0.6524605	0.9549375	0.3849652	0.133609	0.8235335
53	0.7301306	0.927046	0.3315055	0.6952571	0.7492601	0.7204498	0.5391008	0.5479409	0.1085772	0.1490579
54	0.4748944	0.8061566	0.4410544	0.390615	0.524303	0.3681207	0.7457649	0.0583988	0.384263	0.6110552
55										
56										
57						Frequency				
58						Bin	Frequency			
59						0	0			
60						0.1	47			
61						0.2	52			
62						0.3	45			
63						0.4	54			
64						0.5	58			
65						0.6	49			
66						0.7	57			
67						0.8	50			
68						0.9	39			
69						More	49			

Figure 2.1 Random Numbers Generated by Excel

In Figure 2.1 (file T14–1.xls or T14–1.wk1) we generated 500 random numbers by copying the formula

```
Lotus: @RAND
Excel: =RAND()
```

from the cell A5 to the range A5:J54. Figure 2.1 displays the output from Excel. A frequency distribution or histogram of these random numbers is also given in Figure 2.1. By the way, we obtained this frequency distribution by choosing Tools, then Data Analysis, and finally the Histogram command. With Lotus 1-2-3 for Windows you can obtain a frequency distribution with the Range Analyze Distribution command and then create a histogram by using the Lotus graphing capabilities.

Ideally, we would expect each of the intervals of length .1 to contain around .10(500) = 50 random numbers. Each of the intervals of length .10 in column F (0–.1, .1–.2, ... , .9–1) of Figure 2.1 contains between 7.8% and 11.6% of the random numbers. This appears fairly consistent with Property 1 of random numbers. Using a chi-square test we could formally test whether our sample of 500 random numbers was consistent with Property 1. If we generated many (say 1,000,000) random numbers we could be quite sure that each of the above intervals would contain almost exactly 10% of the generated random numbers.

To check whether our random numbers are consistent with Property 2 (the independence property), we determined the fraction of the time that a small (e.g., $\leq .5$) random number was immediately followed by a big (e.g., $>.5$) random number. This happened 51% of the time, which appears to be consistent with the independence property. Note that this is not enough to verify the independence property. It could be that 90% of the time two consecutive small ($\leq .5$) random numbers are followed by a large ($>.5$) random number. This result would violate the independence property. It is difficult to determine whether the spreadsheet's method for generating random numbers satisfies the independence property.

Problems

Group A

2.1 Use the `@RAND` (or `=RAND()`) and Copy commands to generate a set of 100 random numbers.

a What fraction of the random numbers are smaller than .5?

b What fraction of the time is a random number $\leq .5$ followed by a random number $>.5$?

c What fraction of the random numbers are larger than .8?

Group B

2.2 We all hate to bring small change to the store. Using random numbers, we can eliminate the need for change and give the store and customer a fair shake.

a Suppose you buy something that costs $.20. How could you use random numbers (built into the cash register system) to decide whether you should pay $1.00 or nothing? This eliminates the need for change!

b If you bought something for $9.60 how would you use random numbers to eliminate the need for change?

c In the long run, why is this method fair to both the store and the customer?

CHAPTER 3

Using Spreadsheets to Perform Simulations

In this section we show how spreadsheets may be used to perform simulations of situations in which the uncertainty occurs through a **discrete random variable(s)**. For our purposes, a random variable can be considered discrete if it assumes only a finite number of values. To illustrate this method we show how to simulate a simple newsvendor problem on a spreadsheet.

Example 3.1

In August, Walton Bookstore must decide how many of next year's nature calendars should be ordered. Each calendar costs the bookstore $2.00 and is sold for $4.50. After January 1, any unsold calendars are returned to the publisher for a refund of $.75 per calendar. Walton believes that the number of calendars sold by January 1 follows the probability distribution shown in Table 3.1. Walton wishes to maximize the expected net profit from calendar sales. Use simulation to determine how many calendars the bookstore should order in August.

Table 3.1 Demand Distribution for Calendars at Walton Bookstore

No. of Calendars Sold	Probability
100	.30
150	.20
200	.30
250	.15
300	.05

Solution For a given order quantity, we will show how Excel or Lotus can be used to simulate 50 (or any number) trials. Each trial is an independent "replay" of Walton's problem. To illustrate, suppose we wish to estimate expected profit if 200 calendars are ordered. For an order quantity of 200 calendars Figure 3.1 illustrates the results obtained by simulating 50 independent trials. Here's an explanation of how we created Figure 3.1 (file News.xls or News.wk4).

Step 1 In cell A3 we entered 200 (the order quantity) and copied the 200 to the cell range A4:A52.

Step 2 In cell B3 we generate a $U(0, 1)$ random number for the first trial by entering

```
Excel:  =RAND()

Lotus:  @RAND
```

Step 3 We copy this formula from cell B3 to cells B3:B52. This generates 50 independent $U(0, 1)$ random numbers. Note that when we copy `=RAND()` or `@RAND` from B3, the random number in cell B3 changes. This is because any recalculation of the spreadsheet causes the spreadsheet to generate "new" random numbers. To prevent this from happening in Excel or Lotus you can "freeze" the values of your random numbers with the Edit command. First copy the contents of cell range B3:B52 to the clipboard and then use the Paste Special command and Values option to paste the values of the random numbers to the cell range B3:B52. From this point on our 50 random numbers will remain unchanged.

Step 4 In column C we generate the annual demand for calendars for each of 50 trials. Given the probabilities for calendar demand, the correspondence between $U(0, 1)$ random numbers and quantity demanded should be as given in Table 3.2.

	B	C	D	E	F
1			FIGURE 3.1 OQ=200		
2	RN	QUANTDEM	REVENUE	COST	PROFIT
3	0.172714	100	525	400	125
4	0.2266297	100	525	400	125
5	0.2300341	100	525	400	125
6	0.5618113	200	900	400	500
7	0.9430297	250	900	400	500
8	0.6733184	200	900	400	500
9	0.6240083	200	900	400	500
10	0.5756413	200	900	400	500
11	0.2931459	100	525	400	125
12	0.3013666	150	712.5	400	312.5
13	0.4726825	150	712.5	400	312.5
14	0.6576143	200	900	400	500
15	0.7460252	200	900	400	500
16	0.4666824	150	712.5	400	312.5
17	0.5518515	200	900	400	500
18	0.1296437	100	525	400	125
19	0.2233458	100	525	400	125
20	0.5205849	200	900	400	500
21	0.4711761	150	712.5	400	312.5
22	0.8803125	250	900	400	500
23	0.4082341	150	712.5	400	312.5
24	0.844167	250	900	400	500
25	0.3870129	150	712.5	400	312.5
26	0.8215992	250	900	400	500
27	0.429285	150	712.5	400	312.5
28	0.9578531	300	900	400	500
29	0.4349703	150	712.5	400	312.5
30	0.6629866	200	900	400	500
31	0.7252477	200	900	400	500
32	0.5030282	200	900	400	500
33	0.3665663	150	712.5	400	312.5
34	0.9248289	250	900	400	500
35	0.1997494	100	525	400	125
36	0.1426111	100	525	400	125
37	0.1243629	100	525	400	125
38	0.3628507	150	712.5	400	312.5
39	0.0254334	100	525	400	125
40	0.6774485	200	900	400	500
41	0.3399515	150	712.5	400	312.5
42	0.2256545	100	525	400	125
43	0.7068867	200	900	400	500
44	0.9254046	250	900	400	500
45	0.9891658	300	900	400	500
46	0.0171112	100	525	400	125
47	0.261731	100	525	400	125
48	0.1600208	100	525	400	125
49	0.6888955	200	900	400	500
50	0.6855076	200	900	400	500
51	0.6292534	200	900	400	500
52	0.1556201	100	525	400	125
53				AVG	346.25
54				STDEV	163.766

Figure 3.1 Order Quantity = 200

Table 3.2 Correspondence Between Random Numbers and Calendar Demand

Calendars Demanded	Probability of Occurring	Random Number Range
100	.30	$.00 \le RN < .30$
150	.20	$.30 \le RN < .50$
200	.30	$.50 \le RN < .80$
250	.15	$.80 \le RN < .95$
300	.05	$.95 \le RN < 1.00$

From Chapter 2 we know that around 30% of the random numbers generated by the spreadsheet will fall between 0 and .30. Then the correspondence in Table 3.2 ensures that (in the long run) 100 calendars will be demanded in 30% of our trials. Similarly, 20% of the generated random numbers should be between .30 and .50. This will lead to a demand of 150 calendars during 20% of all trials. *The correspondence between the random numbers and the generated demand is the key to the way in which the simulation imitates the real-life situation!* By using the correspondence given in Table 3.2 we can ensure that the computer generates demands with probabilities that agree with the probabilities that we believe represent the real-life situation.

We can incorporate the relationship between the random numbers and quantity demanded by using either an `IF` or `LOOKUP` command. We choose to use an `IF` command. `LOOKUP` tables are discussed, however, in Chapter 14. To generate the calendar demand for our first trial (row 3 of the spreadsheet) we enter the following in cell C3:

```
Excel: =IF(B3<.3,100,IF(B3<.5,150,IF(B3<.8,200,
  IF(B3<.95,250,300))))

Lotus: @IF(B3<.3,100,@IF(B3<.5,150,@IF(B3<.8,200,
  @IF(B3<.95,250,300))))
```

This statement has the following effect: If B3 < .3, set C3 = 100; otherwise check to see whether B3 < .5. If B3 < .5, then set C3 = 150. If B3 < .5 is false, then check whether B3 < .8. If B3 < .8, then set C3 = 200. If B3 < .8 is not true, check whether B3 < .95. If B3 < .95 is true, set C3 = 250. Finally, if C3 < .95 is not true, set C3 = 300. If, for example, the random number in B3 is .76, this statement will set C3 = 200, generating an annual demand of 200 calendars. By the way, we need one ending parentheses for each `IF` in our formula, so we use four ending parentheses.

Step 5 In column D we generate our total revenue. To do this we enter the statement

```
Excel: =4.5*MIN(A3,C3)+.75*MAX(0,A3-C3)

Lotus: 4.5*@MIN(A3,C3)+.75*@MAX(0,A3-C3)
```

in D3. To see why this statement generates total revenue recall that we receive $4.50 for each calendar sold and

Sold Calendars = Minimum (Calendars Ordered, Demand for Calendars)

Thus revenue for calendars sold is generated by `4.5*@MIN(A3,C3)`. Also recall that each leftover calendar generates $0.75 in revenue and

Leftover Calendars = Maximum (0, Calendars Ordered – Calendars Demanded)

Thus `.75*@MAX(0,A3-C3)` generates revenue from leftover calendars. By the way, we could have computed profit with the formula

```
Excel:  =4.5*IF(A3<C3,A3,C3)+.75*IF(A3<C3,0,A3-C3)

Lotus:  4.5*@IF(A3<C3,A3,C3)+.75*@IF(A3<C3,0,A3-C3)
```

Step 6 In E3 we generate cost of ordering calendars by multiplying calendars ordered by 2. Thus we enter in E3

```
Excel:  =2*A3

Lotus:  2*A3
```

Step 7 In F3 we generate Profit = Revenue – Costs by entering

```
Excel:  =D3-E3

Lotus:  +D3-E3
```

Step 8 We now generate 50 trials of our simulation by copying the formulas in range C3:F3 to the range C3:F52.

Step 9 In F53 we find the average profit for our 50 trials by entering

```
Excel:  =AVERAGE(F3:F52)

Lotus:  @AVG(F3..F52)
```

This shows that for our 50 trials an average profit of $346.25 per year was obtained.

Step 10 In F54 we find the sample standard deviation s of the profit for our 50 trials by entering in F54

```
Excel:  =STDEV(F3:F52)

Lotus:  @STDS(F3..F52)
```

We find that $s = 163.77$. In general, if there are n trials and p_i = profit on the ith trial, then s is the square root of the sample variance s^2 given by

$$s^2 = \frac{1}{n-1}\sum_{i=1}^{i=n}(p_i - \bar{p})^2$$

where $\bar{p}$ = average profit in all n trials.

3.1 Finding a Confidence Interval for Expected Profit

From basic statistics it can be shown that for a sample size of at least 30, a 100(1 – α)% confidence interval for the expected profit associated with an order quantity of 200 calendars is given by

$$\bar{p} \pm t_{(\alpha/2,\, n-1)} s/n^{1/2}$$

Here $t_{(\alpha,\, n)}$ is the number x satisfying Prob($t \geq x$) = α when t follows a t-distribution with n degrees of freedom. You can use the Excel `TINV` function to compute $t_{(\alpha,\, n)}$. Entering `TINV(2α,n)` will return the value $t_{(\alpha,\, n)}$. Thus, to compute $t_{(.025,\, 49)}$ we would enter `=TINV(.05,49)`.

We know that $\bar{p}$ = 346.25 and s = 163.77. Entering `=TINV(.05,49)` we find that $t_{(.025,\, 49)}$ = 2.01. Thus our 95% confidence interval for expected profit (given an order quantity of 200 calendars) is

$$346.25 \pm 163.77(2.01)/50^{1/2} \text{ or } \$299.70 < \text{Expected Profit} < \$392.80$$

3.2 How Many Trials Do We Need?

Suppose that we want to use simulation to estimate a parameter (such as expected profit when 200 calendars are ordered) and be accurate within an amount D, a fraction 100(1 – α)% of the time. If the parameter has a standard deviation σ, then the number of trials (n) needed is given by

$$n = z_{\alpha/2}{}^2\sigma^2/D^2$$

where z_α satisfies $P(Z > z_\alpha) = \alpha$, and Z follows a normal distribution with mean 0 and standard deviation 1. The Excel `NORMSINV` function can be used to compute z_α. To compute z_α just enter `=NORMSINV(1-α)`. For example, to compute $z_{.025}$ just enter `=NORMSINV(.975)`, and Excel will return 1.96.

Let σ = standard deviation of annual profit if 200 calendars are ordered. Suppose we wish to estimate expected annual profit within \$10 when 200 calendars are ordered, and we want to be this accurate 95% of the time. If we now assume that the sample standard deviation obtained by our 50 trials (s = 163.77) is a reasonable estimate of σ, we find that the estimated number of trials is given by

$$n = z_{.025}{}^2(163.77)^2/(10)^2 = (1.96)^2(163.77)^2/(10)^2 = 1030.34$$

Thus a total of 1031 trials (or 1031 – 50 = 981 more trials) must be conducted to be 95% sure that our estimate of expected profit is accurate within \$10.

3.3 Determination of the Optimal Order Quantity

To determine the order quantity that maximizes expected profit, we now modify column A of our spreadsheet and change the order quantity to 100, 150, 250, and 300 calendars (to see why we don't consider other possible order quantities see Problem 3.1). We use the same random numbers to evaluate each order quantity. We use the same numbers because generating an additional set of random numbers for each potential order quantity would introduce an additional source of variability into the analysis, which might obscure the difference in profit caused by changing the order quantity. Figures 3.2–3.5 give the results of these simulations. Our point estimate of expected profit for each order quantity is summarized in Table 3.3. For each order quantity we also give (in parentheses) the actual expected profit. For example, for an order quantity of 150

Expected Cost = 2(150) = 300

Expected Revenue for \$4.50 Sales = (.3)(100*4.50) + (.7)(150*4.50) = \$607.50

Expected Revenue for \$0.75 Sales = .3(50*.75) = 11.25

Expected Profit = 607.50 + 11.25 – 300 = \$318.75

Table 3.3 Expected Profit from Newsvendor Simulation

Order Quantity	Expected Profit
100	\$250.00 (\$250)
150	\$318.75 (\$318.75)
200	\$346.25 (\$350)
250	\$313.75 (\$325)
300	\$258.75 (\$271.88)

Thus it appears that ordering 200 calendars will maximize expected profit. By using methods discussed in Section 17.3 of Winston (1994), it can be shown that an order quantity of 200 does maximize expected profit.

Recall, however, that the width of our 95% confidence interval for expected profit for an order quantity of 200 calendars was approximately \$93. This indicates that many more trials should be conducted to "hone in" on the expected profit for each possible order quantity.

	A	B	C	D	E	F
1				FIGURE 3.2 OQ=100		
2	ORDERQUAN	RN	QUANTDEM	REVENUE	COST	PROFIT
3	100	0.172714	100	450	200	250
4	100	0.2266297	100	450	200	250
5	100	0.2300341	100	450	200	250
6	100	0.5618113	200	450	200	250
7	100	0.9430297	250	450	200	250
8	100	0.6733184	200	450	200	250
9	100	0.6240083	200	450	200	250
10	100	0.5756413	200	450	200	250
11	100	0.2931459	100	450	200	250
12	100	0.3013666	150	450	200	250
13	100	0.4726825	150	450	200	250
14	100	0.6576143	200	450	200	250
15	100	0.7460252	200	450	200	250
16	100	0.4666824	150	450	200	250
17	100	0.5518515	200	450	200	250
18	100	0.1296437	100	450	200	250
19	100	0.2233458	100	450	200	250
20	100	0.5205849	200	450	200	250
21	100	0.4711761	150	450	200	250
22	100	0.8803125	250	450	200	250
23	100	0.4082341	150	450	200	250
24	100	0.844167	250	450	200	250
25	100	0.3870129	150	450	200	250
26	100	0.8215992	250	450	200	250
27	100	0.429285	150	450	200	250
28	100	0.9578531	300	450	200	250
29	100	0.4349703	150	450	200	250
30	100	0.6629866	200	450	200	250
31	100	0.7252477	200	450	200	250
32	100	0.5030282	200	450	200	250
33	100	0.3665663	150	450	200	250
34	100	0.9248289	250	450	200	250
35	100	0.1997494	100	450	200	250
36	100	0.1426111	100	450	200	250
37	100	0.1243629	100	450	200	250
38	100	0.3628507	150	450	200	250
39	100	0.0254334	100	450	200	250
40	100	0.6774485	200	450	200	250
41	100	0.3399515	150	450	200	250
42	100	0.2256545	100	450	200	250
43	100	0.7068867	200	450	200	250
44	100	0.9254046	250	450	200	250
45	100	0.9891658	300	450	200	250
46	100	0.0171112	100	450	200	250
47	100	0.261731	100	450	200	250
48	100	0.1600208	100	450	200	250
49	100	0.6888955	200	450	200	250
50	100	0.6855076	200	450	200	250
51	100	0.6292534	200	450	200	250
52	100	0.1556201	100	450	200	250
53					AVG	250
54					STDEV	0

Figure 3.2 Order Quantity = 100

	A	B	C	D	E	F
1				FIGURE 3.3 OQ=150		
2	ORDERQUAN	RN	QUANTDEM	REVENUE	COST	PROFIT
3	150	0.172714	100	487.5	300	187.5
4	150	0.2266297	100	487.5	300	187.5
5	150	0.2300341	100	487.5	300	187.5
6	150	0.5618113	200	675	300	375
7	150	0.9430297	250	675	300	375
8	150	0.6733184	200	675	300	375
9	150	0.6240083	200	675	300	375
10	150	0.5756413	200	675	300	375
11	150	0.2931459	100	487.5	300	187.5
12	150	0.3013666	150	675	300	375
13	150	0.4726825	150	675	300	375
14	150	0.6576143	200	675	300	375
15	150	0.7460252	200	675	300	375
16	150	0.4666824	150	675	300	375
17	150	0.5518515	200	675	300	375
18	150	0.1296437	100	487.5	300	187.5
19	150	0.2233458	100	487.5	300	187.5
20	150	0.5205849	200	675	300	375
21	150	0.4711761	150	675	300	375
22	150	0.8803125	250	675	300	375
23	150	0.4082341	150	675	300	375
24	150	0.844167	250	675	300	375
25	150	0.3870129	150	675	300	375
26	150	0.8215992	250	675	300	375
27	150	0.429285	150	675	300	375
28	150	0.9578531	300	675	300	375
29	150	0.4349703	150	675	300	375
30	150	0.6629866	200	675	300	375
31	150	0.7252477	200	675	300	375
32	150	0.5030282	200	675	300	375
33	150	0.3665663	150	675	300	375
34	150	0.9248289	250	675	300	375
35	150	0.1997494	100	487.5	300	187.5
36	150	0.1426111	100	487.5	300	187.5
37	150	0.1243629	100	487.5	300	187.5
38	150	0.3628507	150	675	300	375
39	150	0.0254334	100	487.5	300	187.5
40	150	0.6774485	200	675	300	375
41	150	0.3399515	150	675	300	375
42	150	0.2256545	100	487.5	300	187.5
43	150	0.7068867	200	675	300	375
44	150	0.9254046	250	675	300	375
45	150	0.9891658	300	675	300	375
46	150	0.0171112	100	487.5	300	187.5
47	150	0.261731	100	487.5	300	187.5
48	150	0.1600208	100	487.5	300	187.5
49	150	0.6888955	200	675	300	375
50	150	0.6855076	200	675	300	375
51	150	0.6292534	200	675	300	375
52	150	0.1556201	100	487.5	300	187.5
53					AVG	318.75
54					STDEV	86.7956

Figure 3.3 Order Quantity = 150

	A	B	C	D	E	F
1				FIGURE 3.4 OQ=250		
2	ORDERQUAN	RN	QUANTDEM	REVENUE	COST	PROFIT
3	250	0.172714	100	562.5	500	62.5
4	250	0.2266297	100	562.5	500	62.5
5	250	0.2300341	100	562.5	500	62.5
6	250	0.5618113	200	937.5	500	437.5
7	250	0.9430297	250	1125	500	625
8	250	0.6733184	200	937.5	500	437.5
9	250	0.6240083	200	937.5	500	437.5
10	250	0.5756413	200	937.5	500	437.5
11	250	0.2931459	100	562.5	500	62.5
12	250	0.3013666	150	750	500	250
13	250	0.4726825	150	750	500	250
14	250	0.6576143	200	937.5	500	437.5
15	250	0.7460252	200	937.5	500	437.5
16	250	0.4666824	150	750	500	250
17	250	0.5518515	200	937.5	500	437.5
18	250	0.1296437	100	562.5	500	62.5
19	250	0.2233458	100	562.5	500	62.5
20	250	0.5205849	200	937.5	500	437.5
21	250	0.4711761	150	750	500	250
22	250	0.8803125	250	1125	500	625
23	250	0.4082341	150	750	500	250
24	250	0.844167	250	1125	500	625
25	250	0.3870129	150	750	500	250
26	250	0.8215992	250	1125	500	625
27	250	0.429285	150	750	500	250
28	250	0.9578531	300	1125	500	625
29	250	0.4349703	150	750	500	250
30	250	0.6629866	200	937.5	500	437.5
31	250	0.7252477	200	937.5	500	437.5
32	250	0.5030282	200	937.5	500	437.5
33	250	0.3665663	150	750	500	250
34	250	0.9248289	250	1125	500	625
35	250	0.1997494	100	562.5	500	62.5
36	250	0.1426111	100	562.5	500	62.5
37	250	0.1243629	100	562.5	500	62.5
38	250	0.3628507	150	750	500	250
39	250	0.0254334	100	562.5	500	62.5
40	250	0.6774485	200	937.5	500	437.5
41	250	0.3399515	150	750	500	250
42	250	0.2256545	100	562.5	500	62.5
43	250	0.7068867	200	937.5	500	437.5
44	250	0.9254046	250	1125	500	625
45	250	0.9891658	300	1125	500	625
46	250	0.0171112	100	562.5	500	62.5
47	250	0.261731	100	562.5	500	62.5
48	250	0.1600208	100	562.5	500	62.5
49	250	0.6888955	200	937.5	500	437.5
50	250	0.6855076	200	937.5	500	437.5
51	250	0.6292534	200	937.5	500	437.5
52	250	0.1556201	100	562.5	500	62.5
53					AVG	313.75
54					STDEV	202.618

Figure 3.4 Order Quantity = 250

	A	B	C	D	E	F
1				FIGURE 3.5 OQ=300		
2	ORDERQUAN	RN	QUANTDEM	REVENUE	COST	PROFIT
3	300	0.172714	100	600	600	0
4	300	0.2266297	100	600	600	0
5	300	0.2300341	100	600	600	0
6	300	0.5618113	200	975	600	375
7	300	0.9430297	250	1162.5	600	562.5
8	300	0.6733184	200	975	600	375
9	300	0.6240083	200	975	600	375
10	300	0.5756413	200	975	600	375
11	300	0.2931459	100	600	600	0
12	300	0.3013666	150	787.5	600	187.5
13	300	0.4726825	150	787.5	600	187.5
14	300	0.6576143	200	975	600	375
15	300	0.7460252	200	975	600	375
16	300	0.4666824	150	787.5	600	187.5
17	300	0.5518515	200	975	600	375
18	300	0.1296437	100	600	600	0
19	300	0.2233458	100	600	600	0
20	300	0.5205849	200	975	600	375
21	300	0.4711761	150	787.5	600	187.5
22	300	0.8803125	250	1162.5	600	562.5
23	300	0.4082341	150	787.5	600	187.5
24	300	0.844167	250	1162.5	600	562.5
25	300	0.3870129	150	787.5	600	187.5
26	300	0.8215992	250	1162.5	600	562.5
27	300	0.429285	150	787.5	600	187.5
28	300	0.9578531	300	1350	600	750
29	300	0.4349703	150	787.5	600	187.5
30	300	0.6629866	200	975	600	375
31	300	0.7252477	200	975	600	375
32	300	0.5030282	200	975	600	375
33	300	0.3665663	150	787.5	600	187.5
34	300	0.9248289	250	1162.5	600	562.5
35	300	0.1997494	100	600	600	0
36	300	0.1426111	100	600	600	0
37	300	0.1243629	100	600	600	0
38	300	0.3628507	150	787.5	600	187.5
39	300	0.0254334	100	600	600	0
40	300	0.6774485	200	975	600	375
41	300	0.3399515	150	787.5	600	187.5
42	300	0.2256545	100	600	600	0
43	300	0.7068867	200	975	600	375
44	300	0.9254046	250	1162.5	600	562.5
45	300	0.9891658	300	1350	600	750
46	300	0.0171112	100	600	600	0
47	300	0.261731	100	600	600	0
48	300	0.1600208	100	600	600	0
49	300	0.6888955	200	975	600	375
50	300	0.6855076	200	975	600	375
51	300	0.6292534	200	975	600	375
52	300	0.1556201	100	600	600	0
53					AVG	258.75
54					STDEV	217.245

Figure 3.5 Order Quantity = 300

Remarks

1. This type of simulation is often called a **Monte Carlo simulation,** because the random number used for each trial is analogous to a spin of the roulette wheel at a casino. Like the spins of a roulette wheel, the random numbers used to generate demands for each trial are independent. The term *Monte Carlo simulation* was coined by mathematicians Stanislaw Ulam and James Von Neumann when they developed computer simulations of nuclear fission, which were used to determine whether an atom bomb was feasible. These simulations were given the code name Monte Carlo.
2. From Chapter 2 we know that all the random numbers we have generated are independent. This fact is necessary for our discussion of confidence intervals and sample size to be valid.
3. By hitting the Recalc (F9) button, you can see that each time a simulation is performed, different results will ensue! This assumes, of course, that the `=RAND()` and `@RAND` formulas have not been frozen.
4. Our formula for a confidence interval yields a confidence interval for the mean annual profit. Note that as the number of trials grows large, the width of the confidence interval for mean profit shrinks to 0. This shows that if we run many iterations we will indeed obtain a good estimate of mean profit.
5. Suppose we want a $100(1 - \alpha)\%$ confidence interval for the profit earned during a particular year. Assuming that profit during a particular year is normally distributed (which is not true in Example 3.1), it can be shown that a $100(1 - \alpha)\%$ confidence interval for the profit earned during a particular year is

$$\overline{p} \pm t_{(\alpha/2,n-1)} s\sqrt{1+(1/n)}$$

3.4 Using Excel Data Tables to Repeat a Simulation

Instead of copying the first trial in our spreadsheet to a 50 row range we can use an Excel data table to "simulate" 50 trials of the newsvendor problem. Figure 3.6 illustrates the use of a data table in simulation. To use the data table you need to first select a table range. The table range should contain two (or more) columns in which the results of the simulation will be placed. In our example the range should consist of two columns and a number of rows equal to the number of desired trials + 1. In the top cell of the second column of the table range, you enter the formula you want to be calculated during each trial of the simulation. We proceed as follows:

	F	G	H	I
1			**Figure 3.6**	
2		Trial	500	
3		1	500	
4		2	312.5	
5		3	500	
6		4	312.5	
7		5	500	
8		6	312.5	
9		7	500	
10		8	312.5	
11		9	312.5	
12		10	312.5	
13		11	500	
14		12	500	
15		13	312.5	
16		14	500	
17		15	125	
18		16	312.5	
19		17	125	
20		18	125	
21		19	500	
22		20	500	
23		21	500	
24		22	500	
25		23	500	
26		24	312.5	
27		25	500	
28		26	500	
29		27	125	
30		28	125	
31		29	125	
32		30	312.5	
33		31	312.5	
34		32	312.5	
35		33	125	
36		34	500	
37		35	125	
38		36	500	
39		37	500	
40		38	125	
41		39	500	
42		40	312.5	
43		41	500	
44		42	500	
45		43	312.5	
46		44	500	
47		45	312.5	
48		46	125	
49		47	500	
50		48	500	
51		49	125	
52		50	500	
53		Average	361.25	

Figure 3.6 One-Way Data Table

Step 1 Begin by choosing G2:H52 to be our table range.

Step 2 In cell H2 we enter the formula for profit (`=D3-E3` or `=F3`).

Step 3 Next we select the cell range G2:H52.

Step 4 Choose the commands Data and Table. The Data Table dialog box will then appear.

Step 5 Use the dialog box to select any blank cell in the spreadsheet (say I1) as your column input cell.

In cells H3:H52 (see Figure 3.6) we obtain the profit for 50 trials of the simulation. For each trial, the computer chooses a value of `=RAND()` (the values of `=RAND()` for each trial are independent). Basically, what the data table does in this situation is recalculate all cells in the spreadsheet that influence the formula in H2 50 times. During the *i*th time the formula in H2 is computed, the *i*th value in column G is used (this is because we selected I1 as a column input cell, not a row input cell) as the value in the input cell (I1). Since cell I1 is not referred to in cell H2, this implies that `=D3-E3` will be calculated 50 times. Each time a different value of `=RAND()` is used. By using the command `@AVG(H3:H52)`, we see that the average profit for our 50 trials is $361.25. Of course, this method will only work if the random number in B3 has not been frozen.

	J	K	L	M	N	O	P	Q
1			**Figure 3.7**					
2		125	100	150	200	250	300	
3		1	250	375	312.5	437.5	187.5	
4		2	250	187.5	500	62.5	375	
5		3	250	375	500	625	562.5	
6		4	250	187.5	500	62.5	562.5	
7		5	250	187.5	500	250	750	
8		6	250	375	500	62.5	187.5	
9		7	250	375	500	625	375	
10		8	250	187.5	500	62.5	375	
11		9	250	187.5	312.5	62.5	375	
12		10	250	187.5	125	62.5	187.5	
13		11	250	375	125	437.5	0	
14		12	250	187.5	312.5	625	375	
15		13	250	375	125	62.5	375	
16		14	250	187.5	312.5	437.5	0	
17		15	250	375	500	437.5	187.5	
18		16	250	375	500	62.5	562.5	
19		17	250	375	500	62.5	562.5	
20		18	250	187.5	125	250	187.5	
21		19	250	187.5	500	437.5	0	
22		20	250	187.5	312.5	437.5	0	
23		21	250	375	500	437.5	0	
24		22	250	187.5	500	62.5	375	
25		23	250	375	125	625	750	
26		24	250	375	312.5	250	562.5	
27		25	250	375	500	625	562.5	
28		26	250	375	312.5	62.5	187.5	
29		27	250	187.5	125	437.5	0	
30		28	250	187.5	312.5	62.5	562.5	
31		29	250	375	500	625	187.5	
32		30	250	375	312.5	625	375	
33		31	250	375	312.5	250	0	
34		32	250	187.5	500	625	0	
35		33	250	375	125	62.5	562.5	
36		34	250	375	312.5	62.5	187.5	
37		35	250	375	500	62.5	187.5	
38		36	250	187.5	500	250	375	
39		37	250	375	500	437.5	187.5	
40		38	250	375	500	62.5	0	
41		39	250	375	500	625	0	
42		40	250	187.5	500	62.5	0	
43		41	250	375	125	62.5	0	
44		42	250	187.5	500	437.5	187.5	
45		43	250	375	500	437.5	187.5	
46		44	250	375	500	437.5	187.5	
47		45	250	375	500	625	375	
48		46	250	375	312.5	250	375	
49		47	250	187.5	125	62.5	562.5	
50		48	250	187.5	500	62.5	0	
51		49	250	375	500	250	187.5	
52		50	250	187.5	500	437.5	375	
53		Average	250	296.25	387.5	298.75	273.75	

Figure 3.7 Two-Way Data Table

By using a two-way data table we can simultaneously simulate the profit for each possible order quantity. A two-way data table enables us to evaluate a formula for various values that are input into two different cells. To use a two-way data table to simulate n trials and s order quantities we need a range containing $n + 1$ rows and $s + 1$ columns (see Figure 3.7). We proceed as follows:

Step 1 Choose the table range K2:P52.

Step 2 In the upper left cell of the range (K2) we enter the formula that we want to evaluate (`=D3-E3` or `=F3`).

Step 3 Enter the trial numbers in cells K3:K52 and the order quantities in L2:P2.

Step 4 Select the cell range K2:P52.

Step 5 Define the column input cell to be any blank cell in the spreadsheet (we choose Z1) and the row input cell to be the cell in row 3 of the spreadsheet where the order quantity is located (A3).

For each cell in the data table range that is not in the first row or column of the range, the spreadsheet recalculates the formula in K2, inputting a number in the first column of the table range into the column input cell and inputting a number in the first row of the table range into the row input cell. For instance, in cell M4 Excel will compute profit inputting a 2 into Z1 and a 150 into A3. Thus each cell in the third column of the table will contain a profit value generated for an order quantity of 150. Since the value of the column input cell does not influence profit, we will therefore obtain the profit for each order quantity for 50 independent trials.

As before, the simulation indicates that expected profit is maximized by ordering 200 calendars.

Remark

If you are using Excel data tables to run simulations you should change the Calculation option to Automatic except for tables. Otherwise your data tables will continually recalculate whenever you change any part of your spreadsheet. This is very slow!

3.5 Using a Lotus What-If Table to Repeat a Simulation

Instead of copying the first trial in our spreadsheet to a 50-row range, we can use a Lotus what-if table to "simulate" 50 trials of the newsvendor problem. Figure 3.6 illustrates the use of a what-if table in simulation. To use the what-if table you need to first select a table range. The table range should contain two (or more) columns in which the results of the simulation will be placed. In our example the range should consist of two columns and a number of rows equal to the number of desired trials + 1. We proceed as follows:

Step 1 Choose G2..H52 to be our table range.

Step 2 In the top cell of the second column of the table range you enter the formula you want to be calculated during each trial of the simulation. Thus in cell H2 we enter the formula for profit (`+D3-E3` or `+F3`).

Step 3 Select the cell range G2..H52.

Step 4 Choose the commands Range Analyze and What-If Table. The What-If Table dialog box will then appear.

Step 5 Use the dialog box to indicate that you are using a 1 variable what-if table.

Step 6 Next select any blank cell in the spreadsheet (say I1) as input cell 1.

In cells H3..H52 (see Figure 3.6) we obtain the profit for 50 trials of the simulation. For each trial, the computer chooses a value of `@RAND` (the values of `@RAND` for each trial are independent). Basically, in this situation the what-if table recalculates all cells in the spreadsheet that influence the formula in H2 50 times. During the *i*th time the formula in H2 is computed, the *i*th value in column G is input into cell I1. Since cell I1 is not referred to in cell H2, this implies that `+D3-E3` will be calculated 50 times. Each time a different value of `@RAND` is used. By using the command `@AVG(H3..H52)` we see that the average profit for our 50 trials is $361.25. Of course, this method will only work if the random number in B3 has not been frozen.

By using a what-if table with two variables we can simultaneously simulate the profit for each possible order quantity. A what-if table with two variables enables us to evaluate a formula for various values that are input into two different cells. To use a what-if table to simulate n trials and s order quantities we need a range containing $n + 1$ rows and $s + 1$ columns (see Figure 3.7). We proceed as follows:

Step 1 Choose the table range K2..P52.

Step 2 In the upper left cell of the range (K2) we enter the formula that we want to evaluate (`+D3-E3` or `+F3`).

Step 3 Enter the trial numbers in cells K3..K52 and the order quantities in L2..P2.

Step 4 Select the cell range K2..P52.

Step 5 Define Input Cell 1 to be any blank cell in the spreadsheet (we choose Z1) and Input Cell 2 to be the cell in row 3 of the spreadsheet where the order quantity is located (A3).

For each cell in the what-if table range that is not in the first row or column of the range, the spreadsheet recalculates the formula in K2, inputting a number in the first column of the table range into Input Cell 1 and inputting a number in the first row of the table range into Input Cell 2. For instance in cell M4 Lotus will compute profit, inputting 2 into Z1 and 150 into A3. Thus each cell in the third column of the table will contain a profit value generated for an order quantity of 150. Since the value of Input Cell 1 does not influence profit, we will therefore obtain the profit for each order quantity for 50 independent trials.

As before, the simulation indicates that expected profit is maximized by ordering 200 calendars.

3.6 Performing the Newsvendor Simulation with the Excel Random Number Generator

Excel has the built-in capability to sample from several distributions, including discrete random variables and normal random variables. Figure 3.8 (file News2.xls) shows how to use the Excel random number generator to simulate Example 3.1 for an order quantity of 200. To create this spreadsheet we delete column B from the file News.xls (column B contained the random numbers). Then in cell B3 we use the Excel Random Number Generation option to directly generate the demand for each of our 50 trials. Before doing this we need to enter the possible demands and their respective probabilities in the spreadsheet. We do this in the range F6:G10. Then we proceed as follows:

Step 1 Begin by moving to the cell where you want the first random number to be entered (cell B3). Then from the Tools Data Analysis menu select Random Number Generation.

Step 2 Enter 1 in the Number of Variables box. This indicates that the demands will be placed in a single column.

Step 3 Enter 50 in the Number of Random Numbers box. This indicates that Excel will generate 50 (independent) demands.

Step 4 Select Discrete from the drop-down list box. This indicates that Excel will be generating observations from a discrete (i.e., one that assumes a finite number of values) random variable.

Step 5 Input the range F6:G10 as the value and probability range. This tells Excel that a demand of 100 should occur 30% of the time, a demand of 150 20% of the time, etc.

Step 6 Select the Output Range option and enter the output range B3:B52 to indicate where the random demands should be generated.

Step 7 Hit OK and your 50 demands will be generated!

	A	B	C	D	E	F	G
1	**Figure 3.8 Newsvendor Simulation with Excel Random Number Generator**						
2	ORDERQUAN	QUANDEM	REVENUE	COST	PROFIT		
3	200	100	525	400	125		
4	200	100	525	400	125		
5	200	250	900	400	500		PROB
6	200	150	712.5	400	312.5	100	0.3
7	200	100	525	400	125	150	0.2
8	200	250	900	400	500	200	0.3
9	200	100	525	400	125	250	0.15
10	200	100	525	400	125	300	0.05
11	200	200	900	400	500		
12	200	100	525	400	125		
13	200	100	525	400	125		
14	200	150	712.5	400	312.5		
15	200	100	525	400	125		
16	200	150	712.5	400	312.5		
17	200	300	900	400	500		
18	200	250	900	400	500		
19	200	150	712.5	400	312.5		
20	200	300	900	400	500		
21	200	200	900	400	500		
22	200	100	525	400	125		
23	200	200	900	400	500		
24	200	200	900	400	500		
25	200	150	712.5	400	312.5		
26	200	300	900	400	500		
27	200	200	900	400	500		
28	200	100	525	400	125		
29	200	150	712.5	400	312.5		
30	200	200	900	400	500		
31	200	200	900	400	500		
32	200	200	900	400	500		
33	200	250	900	400	500		
34	200	200	900	400	500		
35	200	100	525	400	125		
36	200	150	712.5	400	312.5		
37	200	100	525	400	125		
38	200	100	525	400	125		
39	200	150	712.5	400	312.5		
40	200	100	525	400	125		
41	200	150	712.5	400	312.5		
42	200	200	900	400	500		
43	200	200	900	400	500		
44	200	200	900	400	500		
45	200	300	900	400	500		
46	200	100	525	400	125		
47	200	200	900	400	500		
48	200	250	900	400	500		
49	200	100	525	400	125		
50	200	100	525	400	125		
51	200	100	525	400	125		
52	200	200	900	400	500		
53				AVGPROF=	331.25		
54				STDDEVPR=	170.463		

Figure 3.8 Newsvendor Simulation with Excel Random Number Generator

Remark

If you use the Random Number Generation option to generate demands and wish to use a data table to compare profits for different order quantities, you should copy the generated demands down the left column of your data table range. The order quantities should be entered across the top of your table. Then choose cell A3 as the row input cell and cell B3 as your column input cell.

Problems

Group A

3.1 Explain why expected profit must be maximized by ordering a quantity equal to some possible demand for calendars. Hint: If this is not the case, then some order quantity, such as 190 calendars, must maximize expected profit. If ordering 190 calendars maximizes expected profit, then it must yield a higher expected profit than an order size of 150. But then an order of 200 calendars must also yield a larger expected profit than 190 calendars. This contradicts the assumed optimality of ordering 190 calendars!

3.2 In August 1997, a car dealer is trying to determine how many 1998 cars should be ordered. Each car ordered in August 1997 costs $10,000. The demand for the dealer's 1998 models has the probability distribution shown in Table 3.4. Each car sells for $15,000. If demand for 1998 cars exceeds the number of cars ordered in August, the dealer must reorder at a cost of $12,000 per car. Excess cars may be disposed of at $9000 per car. Use simulation to determine how many cars should be ordered in August. For your optimal order quantity find a 95% confidence interval for expected profit.

Table 3.4

No. of Cars Demanded	Probability
20	.30
25	.15
30	.15
35	.20
40	.20

3.3 Suppose that Dalton receives no money for the first 50 excess calendars returned, but still receives $0.75 for each other calendar returned. Does this change the optimal order quantity?

3.4 A TSB (Tax Saver Benefit plan) allows you to put money into an account at the beginning of the calendar year that may be used for medical expenses. This amount is not subject to federal tax (hence the phrase TSB). As you pay medical expenses during the year, you are reimbursed by the administrator of the TSB, until the TSB account is exhausted. The catch is, however, any money left in the TSB at the end of the year is lost to you. You estimate that it is equally likely that your medical expenses for next year will be $3000, $4000, $5000, $6000, or $7000. Your federal income tax rate is 40%. Assume your annual salary is $50,000.

a How much should you put in a TSB? Consider both expected disposable income and the standard deviation of disposable income in your answer.

Hint: Your simulation will indicate that two options have nearly the same expected disposable income.

b Does your annual salary influence the correct decision?

CHAPTER 4

An Introduction to @RISK

@RISK is a popular spreadsheet add-in that makes it much easier to perform spreadsheet simulations. The following three features of @RISK make simulation much easier:

1 @RISK contains functions that make it easier to generate observations from most important random variables. For example, entering

```
Excel:  =RISKNORMAL(10,2)
Lotus:  @RISKNORMAL(10,2)
```

in a cell will generate an observation from a normal random variable with mean 10 and standard deviation 2.

2 You may choose any cell(s) in your simulation as an output cell. Then tell @RISK how many times (this is the number of iterations) you want to simulate the spreadsheet. @RISK will "bookkeep" statistics for the chosen cell(s) and also save a histogram (of the values obtained in the cell for all iterations).

3 With the Lotus @RISK `@RISKSIMTABLE` or Excel @RISK `=RISKSIMTABLE` command you may have @RISK simulate the spreadsheet several times, each time inputting a different value into a desired cell. Thus entering the command

```
Excel:  =RISKSIMTABLE({100,150,200,250,300})
Lotus:  @RISKSIMTABLE(100,150,200,250,300)
```

into cell A2 would tell @RISK to perform a simulation with the desired number of iterations five times. In the first simulation 100 would be entered into A2; in the second simulation 150 will be entered into A2, etc. By a single iteration of a simulation we mean evaluating the entire spreadsheet after @RISK has sampled from the relevant random variables and generated random values for all cells that are uncertain in value.

4.1 Simulating the Newsvendor Example with @RISK

We now show how to simulate Example 3.1 with @RISK. The spreadsheet and statistical results are shown in Figure 4.1 (file Newsboy.wk4 or Newsboy.xls). To begin we open @RISK by double clicking on the @RISK icon.

Setting Up the Spreadsheet

In this and all subsequent simulations we will follow the practice of shading our inputs. This will make it easy to see how a sensitivity analysis can be conducted by changing one or more inputs to our simulation. In our simulation spreadsheet models our inputs will include probability distributions (such as those for calendar demand) as well as parameters (such as sales prices and cost per calendar ordered). We now set up our @RISK model to simulate Example 3.1.

Step 1 Begin by entering the sales price, salvage value, and unit purchase cost in the cell range C3:C5 (Figure 4.1a). Then in cell C1 we will set up our spreadsheet so that @RISK can run a simulation for each possible order quantity (100, 150, 200, 250, 300). To do this we enter in cell C1 the formula

```
Excel: =RISKSIMTABLE({100,150,200,250,300})
Lotus: @RISKSIMTABLE(100,150,200,250,300)
```

If we tell @RISK to run five simulations (we will see how to do this shortly), then during the first simulation the value 100 will be placed in cell C1; during the second simulation the value 150 will be placed in cell C1; etc. The first `SIMTABLE` argument will show in the cell. You can use as many `SIMTABLE` functions as you desire in a single worksheet. If more than one `RISKSIMTABLE` function is used then on the first simulation each `RISKSIMTABLE` function assumes the first value in the statement. In the second simulation each `RISKSIMTABLE` function assumes the second value in the statement, etc.

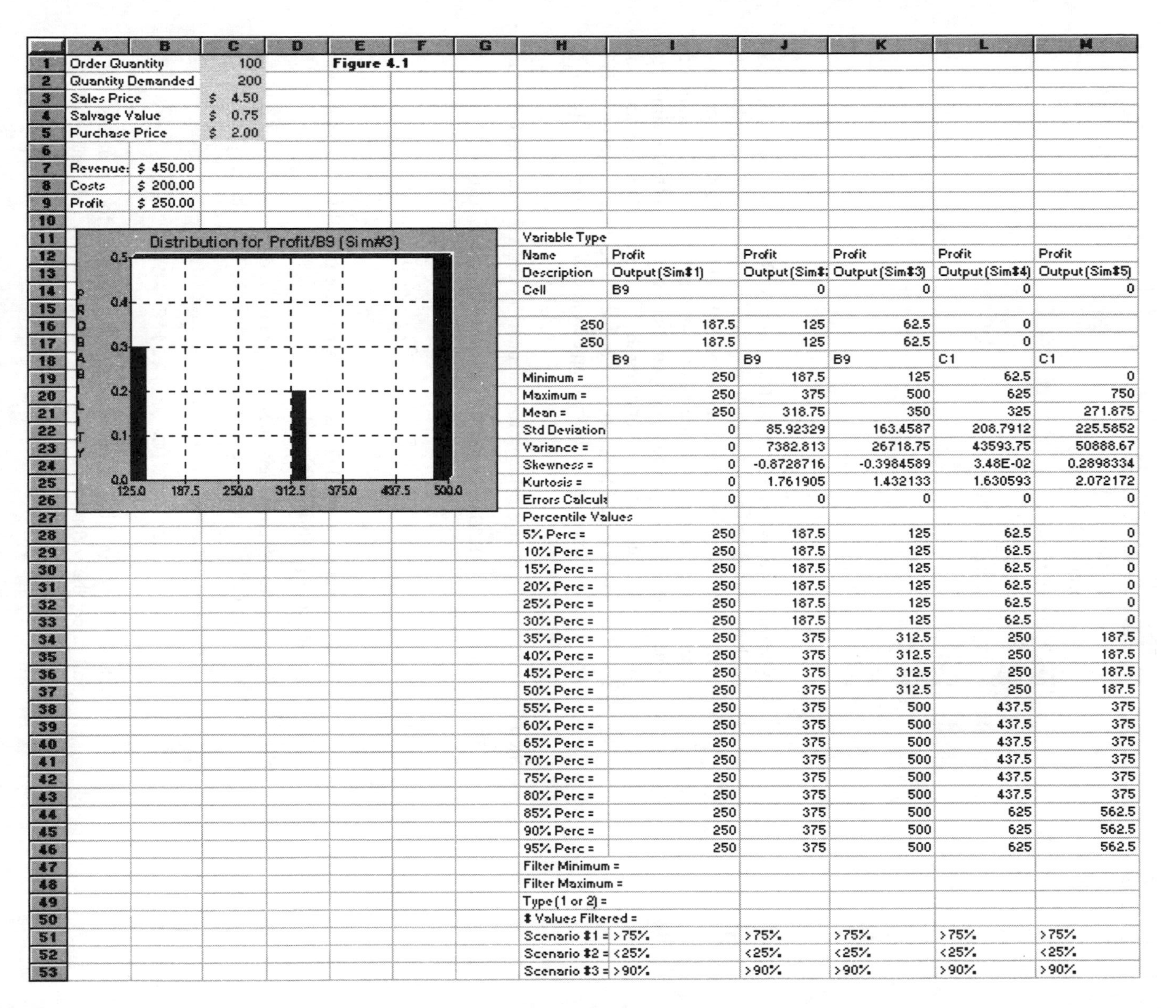

	A	B	C	D	E	F	G	H	I	J	K	L	M
1	Order Quantity		100		Figure 4.1								
2	Quantity Demanded		200										
3	Sales Price		$ 4.50										
4	Salvage Value		$ 0.75										
5	Purchase Price		$ 2.00										
6													
7	Revenue:	$ 450.00											
8	Costs	$ 200.00											
9	Profit	$ 250.00											
10													
11								Variable Type					
12								Name	Profit	Profit	Profit	Profit	Profit
13								Description	Output (Sim#1)	Output (Sim#:	Output (Sim#3)	Output (Sim#4)	Output (Sim#5)
14								Cell	B9	0	0	0	0
15													
16								250	187.5	125	62.5	0	
17								250	187.5	125	62.5	0	
18									B9	B9	B9	C1	C1
19								Minimum =	250	187.5	125	62.5	0
20								Maximum =	250	375	500	625	750
21								Mean =	250	318.75	350	325	271.875
22								Std Deviation	0	85.92329	163.4587	208.7912	225.5852
23								Variance =	0	7382.813	26718.75	43593.75	50888.67
24								Skewness =	0	-0.8728716	-0.3984589	3.48E-02	0.2898334
25								Kurtosis =	0	1.761905	1.432133	1.630593	2.072172
26								Errors Calculs	0	0	0	0	0
27								Percentile Values					
28								5% Perc =	250	187.5	125	62.5	0
29								10% Perc =	250	187.5	125	62.5	0
30								15% Perc =	250	187.5	125	62.5	0
31								20% Perc =	250	187.5	125	62.5	0
32								25% Perc =	250	187.5	125	62.5	0
33								30% Perc =	250	187.5	125	62.5	0
34								35% Perc =	250	375	312.5	250	187.5
35								40% Perc =	250	375	312.5	250	187.5
36								45% Perc =	250	375	312.5	250	187.5
37								50% Perc =	250	375	312.5	250	187.5
38								55% Perc =	250	375	500	437.5	375
39								60% Perc =	250	375	500	437.5	375
40								65% Perc =	250	375	500	437.5	375
41								70% Perc =	250	375	500	437.5	375
42								75% Perc =	250	375	500	437.5	375
43								80% Perc =	250	375	500	437.5	375
44								85% Perc =	250	375	500	625	562.5
45								90% Perc =	250	375	500	625	562.5
46								95% Perc =	250	375	500	625	562.5
47								Filter Minimum =					
48								Filter Maximum =					
49								Type (1 or 2) =					
50								# Values Filtered =					
51								Scenario #1 =	>75%	>75%	>75%	>75%	>75%
52								Scenario #2 =	<25%	<25%	<25%	<25%	<25%
53								Scenario #3 =	>90%	>90%	>90%	>90%	>90%

Figure 4.1 Newsvendor Simulation with @RISK: Discrete Demand

Figure 4.1a

	A	B	C	D	E
1	Order Quantity		100		Figure 4.1
2	Quantity Demanded		200		
3	Sales Price		$ 4.50		
4	Salvage Value		$ 0.75		
5	Purchase Price		$ 2.00		
6					
7	Revenues	$ 450.00			
8	Costs	$ 200.00			
9	Profit	$ 250.00			

When using the `=RISKSIMTABLE` option, @RISK uses the same set of random numbers to evaluate each alternative. Thus, when comparing the profit earned with an order quantity of 200 to the profit earned with an order quantity of, say, 300, you can be sure that during each iteration the same demands were generated. If you chose to run a different simulation for each order quantity without using the `=RISKSIMTABLE` feature, different demands would have been generated for each order quantity. This would make it more difficult to compare the effectiveness of different ordering policies.

Step 2 Generate the annual demand for calendars in cell C2 with the formula

```
Excel: =RISKDISCRETE({100,150,200,250,300},
{.3,.2,.3,.15,.05})

Lotus: @RISKDISCRETE(100,150,200,250,
300,.3,.2,.3,.15,.05)
```

These formulas tell @RISK to enter a number in cell C2 according to the probabilities given in Table 3.1.

Step 3 We compute the costs, revenues, and profit for each iteration of our simulation. In cell B7 we compute revenue with the formula

```
Excel: =C3*MIN(C2,C1)+C4*MAX(0,C1-C2)

Lotus: +C3*@MIN(C2,C1)+C4*@MAX(0,C1-C2)
```

The first term in this formula computes the revenue for calendars sold at full price; the second term computes the revenue for calendars sold at a discount.

Step 4 We compute ordering costs in cell B8 with the formula

```
Excel: =C5*C1
Lotus: +C5*C1
```

Step 5 We compute profit in cell B9 with the formula

```
Excel: =B7-B8
Lotus: +B7-B8
```

Running the Simulation

We are now ready to run our first @RISK simulation! To run the simulation proceed as follows:

Step 1 We must tell @RISK the spreadsheet cells for which data should be collected. These are called **output cells**. We only want to keep track of profit, so we move the cursor to cell B9 and click on the Output icon (fourth icon from left; the icon with one arrow). This will add cell B9 as an output cell. If you want to keep track of a range of cells, preselect the range and then click on the Output icon. @RISK will choose a name for an output cell (or range of cells) based on the headings in your spreadsheet. If you want to change the name of an output range or delete an output range, click on the List icon (fifth from left; the icon with two arrows). To change the name of an output range, simply type over the output range. To delete an output range name, simply click on the Delete button.

Step 2 Click on the Simulation Setting icon (third icon from left; icon has a probability distribution and a square). A Simulation Setting dialog box will appear. Many settings of the simulation can be adjusted, but we will only discuss the crucial settings. We begin by setting the number of iterations to 100. This means that @RISK will recalc the spreadsheet 100 times and keep track of the value in cell B9 for each iteration. Next, we change the number of simulations to 5. In conjunction with our `SIMTABLE` entry in cell C1, this ensures that @RISK will run five simulations of 100 iterations. The first 100 iterations will use an order quantity of 100, the next 100 iterations will use an order quantity of 150, etc. Finally, we change the Standard Recalc option to Monte Carlo. This ensures that if you hit F9 the spreadsheet will recalc. If you leave the Recalc option on Expected Value, then in each cell where you input a probability distribution, @RISK will enter the distribution's expected value (if the random variable is continuous). If the random variable is discrete, then the Expected Value option will result in the possible value of the random variable closest to the actual expected value being displayed in the cell. The True Expected Value option will always cause a random variable's actual expected value to be displayed in a cell. To illustrate, suppose we enter

```
=RISKDISCRETE({100,150,200},{.3,..3,.4})
```

in a cell. If we choose the Monte Carlo option and repeatedly hit F9, we will see a 100 30% of the time, a 150 30% of the time, and a 200 40% of the time. The expected value of this random variable is

$$.3(100) + .3(150) + .4(200) = 155$$

Thus, if we choose the Expected Value option, 150 will be displayed (since 150 is closer to 155 than 100 or 200). If we choose the True Expected Value option, 155 will be displayed.

Unless prompted otherwise, @RISK will keep statistics for all cells containing probability distributions. If your simulation contains many probability distributions, this will create huge files and slow things down. Therefore, it may be advantageous to turn off the Collect Distribution Samples option. If you want to take advantage of tornado graphs and scenario analysis (see Chapter 6), however, you must collect distribution samples.

You can also accelerate your simulation by turning off the Monitor Convergence option. The Monitor Convergence option lets you know how much the mean value of your output cell changes as the simulation progresses. If, for example, the mean profit in your simulation changes by only 1% from Iteration 400 to Iteration 500, you can be fairly confident that you have run enough iterations of your simulation.

Leaving on the Update Display option enables you to see each iteration of the simulation. This is instructive for beginners but slows down the simulation considerably.

Step 3 We are now ready to run our simulation. Click on the Simulate icon (sixth from left; the icon with a probability distribution only), and your simulation will start. The progress of the simulation is monitored in the lower left corner of your screen.

Step 4 When the simulation is completed, you will see a screen with two sets of results: for Simulation #1, a summary (at top of screen) and a more complete statistics section (at bottom of screen). You can use the clipboard to copy your summary or statistics results into your worksheet, where they may be edited. If you want to go back to your original worksheet, click on the Hide icon.

Figure 4.1b

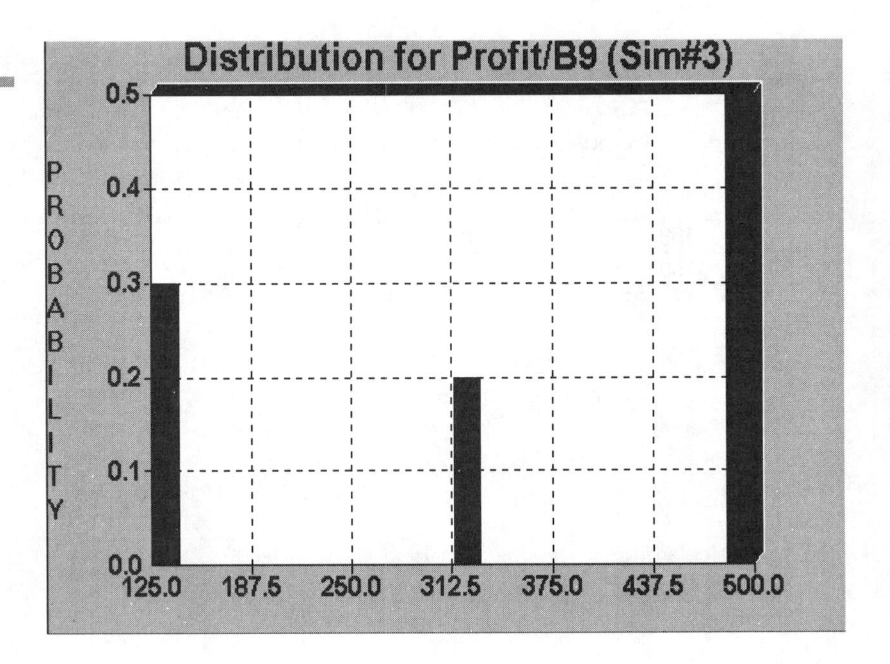

Step 5 You can page through your five simulations by clicking on Sensitivity Next Simulation.

Step 6 If you want to graph the results for any output cell or the probability distribution associated with any input cell, highlight that cell in the summary section of the results and then click on Graph. For Simulation #3 (corresponding to an order quantity of 200 calendars), we obtain the histogram in Figure 4.1b. Any graph created by @RISK can be pasted into your worksheet by using the Clipboard.

Step 7 Assuming that you have just pasted your histogram into your worksheet, click on the Results icon (last icon; the one with a person pointing at a blackboard) to return to the results summary. Now we will merge the statistical summaries from all five simulations and paste them into our worksheet. To do this maximize the Summary window and click on Sensitivity and Merge Simulation. Now click on the statistics section of the results and use the Clipboard to paste the statistical summary into your worksheet.

Step 8 If you wish, you may use the Save icon (second from left; the icon with an arrow pointing into disk) to save the results of your simulation. At a later time, the Open icon (first from left; the icon with an arrow pointing away from disk) may be used to open the results of a saved simulation.

As we will soon see, there is a lot more you can do with @RISK, but for now we have seen enough!

Variable Type					
Name	Profit	Profit	Profit	Profit	Profit
Description	Output (Sim#1)	Output (Sim#2	Output (Sim#3)	Output (Sim#4)	Output (Sim#5)
Cell	B9	0	0	0	0
250	187.5	125	62.5	0	
250	187.5	125	62.5	0	
	B9	B9	B9	C1	C1
Minimum =	250	187.5	125	62.5	0
Maximum =	250	375	500	625	750
Mean =	250	318.75	350	325	271.875
Std Deviation =	0	85.92329	163.4587	208.7912	225.5852
Variance =	0	7382.813	26718.75	43593.75	50888.67
Skewness =	0	-0.8728716	-0.3984589	3.48E-02	0.2898334
Kurtosis =	0	1.761905	1.432133	1.630593	2.072172
Errors Calculat	0	0	0	0	0
Percentile Values					
5% Perc =	250	187.5	125	62.5	0
10% Perc =	250	187.5	125	62.5	0
15% Perc =	250	187.5	125	62.5	0
20% Perc =	250	187.5	125	62.5	0
25% Perc =	250	187.5	125	62.5	0
30% Perc =	250	187.5	125	62.5	0
35% Perc =	250	375	312.5	250	187.5
40% Perc =	250	375	312.5	250	187.5
45% Perc =	250	375	312.5	250	187.5
50% Perc =	250	375	312.5	250	187.5
55% Perc =	250	375	500	437.5	375
60% Perc =	250	375	500	437.5	375
65% Perc =	250	375	500	437.5	375
70% Perc =	250	375	500	437.5	375
75% Perc =	250	375	500	437.5	375
80% Perc =	250	375	500	437.5	375
85% Perc =	250	375	500	625	562.5
90% Perc =	250	375	500	625	562.5
95% Perc =	250	375	500	625	562.5
Filter Minimum =					
Filter Maximum =					
Type (1 or 2) =					
# Values Filtered =					
Scenario #1 =	>75%	>75%	>75%	>75%	>75%
Scenario #2 =	<25%	<25%	<25%	<25%	<25%
Scenario #3 =	>90%	>90%	>90%	>90%	>90%

Figure 4.1c

4.2 Explanation of Statistical Results

To explain the statistical information given by @RISK, we focus on the results for Simulation #3 (order quantity = 200). From column K of Figure 4.1c we obtain the following information:

1 Expected/Mean Result: The average profit for the 100 iterations run for an order quantity of 200 was $350.

2 Maximum Result: For example, the largest observed profit for an order quantity of 200 was $500.

3 Minimum Result: For an order quantity of 200 the smallest observed profit was $125.

4 Standard Deviation: For the 100 iterations run with an order quantity of 200, the sample standard deviation of the profits about the mean profit was 163.46. Of course, the variance is simply the square of the standard deviation. The variance measures the average squared deviation of profits about the mean.

5 Skewness: Skewness is a measure of symmetry. A normal distribution would have a skewness of 0. For an order quantity of 200 the skewness is –.398. Distributions with long positive tails extending to the right are positively skewed, while distributions with long negative tails extending to the left are negatively skewed.

6 Percentile Probabilities: For a discrete random variable such as profit, percentile probability tells us, for example, that for an order quantity of 200 there is at least a 55% chance that profit will be $500 or less and there is at least a 35% chance that profit will be $312.50 or less. If profit were a continuous random variable, then the printout would tell us that (according to the simulation results) there is exactly a 55% chance that profit will be $500 or less when the order quantity equals 200.

We caution the reader that the statistics obtained in any @RISK simulation are estimates of the population parameters. No matter how many iterations we run, we will never know the exact value of population parameters, but the more iterations we run, the higher the probability that our estimate of a population parameter is within a desired amount of the actual population value.

4.3 Conclusions

As before, we see that ordering 200 calendars maximizes expected profit. Note that the profit for ordering 200 calendars has a much higher standard deviation than the profit for ordering 150 calendars. Thus a decision maker who abhors risk might be willing to "trade" $31.25 of expected profit for a reduction in the standard deviation of profit from $163 to $86. In other words, if we are sufficiently risk averse (see Section 12.2 of Winston 1994), then ordering 150 calendars may be a better option than ordering 200 calendars.

By varying our inputs such as sales price and cost per calendar, we may determine the sensitivity of the optimal order quantity to our inputs.

CHAPTER 5

Generating Normal Random Variables

Many continuous random variables that are symmetric about their mean follow (or may be closely approximated by) a normal random variable. In this section we will learn how to use @RISK to generate normal random variables.

5.1 Simulating Normal Demand with @RISK

Let us now suppose that we believe that the demand for next year's calendars will follow a normal distribution with a mean of 200 and a standard deviation of 30. To simulate normal demand with @RISK all we change from Figure 4.1 is the formula for demand entered in cell C2 (see Figure 5.1 and file Normrisk.wk4 or Normrisk.xls). We simply type

```
Excel: =RISKNORMAL(200,30)

Lotus: @RISKNORMAL(200,30)
```

in cell C2 to ensure that demand follows a normal distribution with mean 200 and standard deviation 30 (see Figure 5.1a, b). If you are worried about the fact that there is a (small) probability of a negative demand occurring you could enter the formula

```
Excel: =RISKTNORMAL(200,30,0,1000)

Lotus: @RISKTNORMAL(200,30,0,1000)
```

in cell C2. This generates demands that follow a **truncated normal** distribution. That

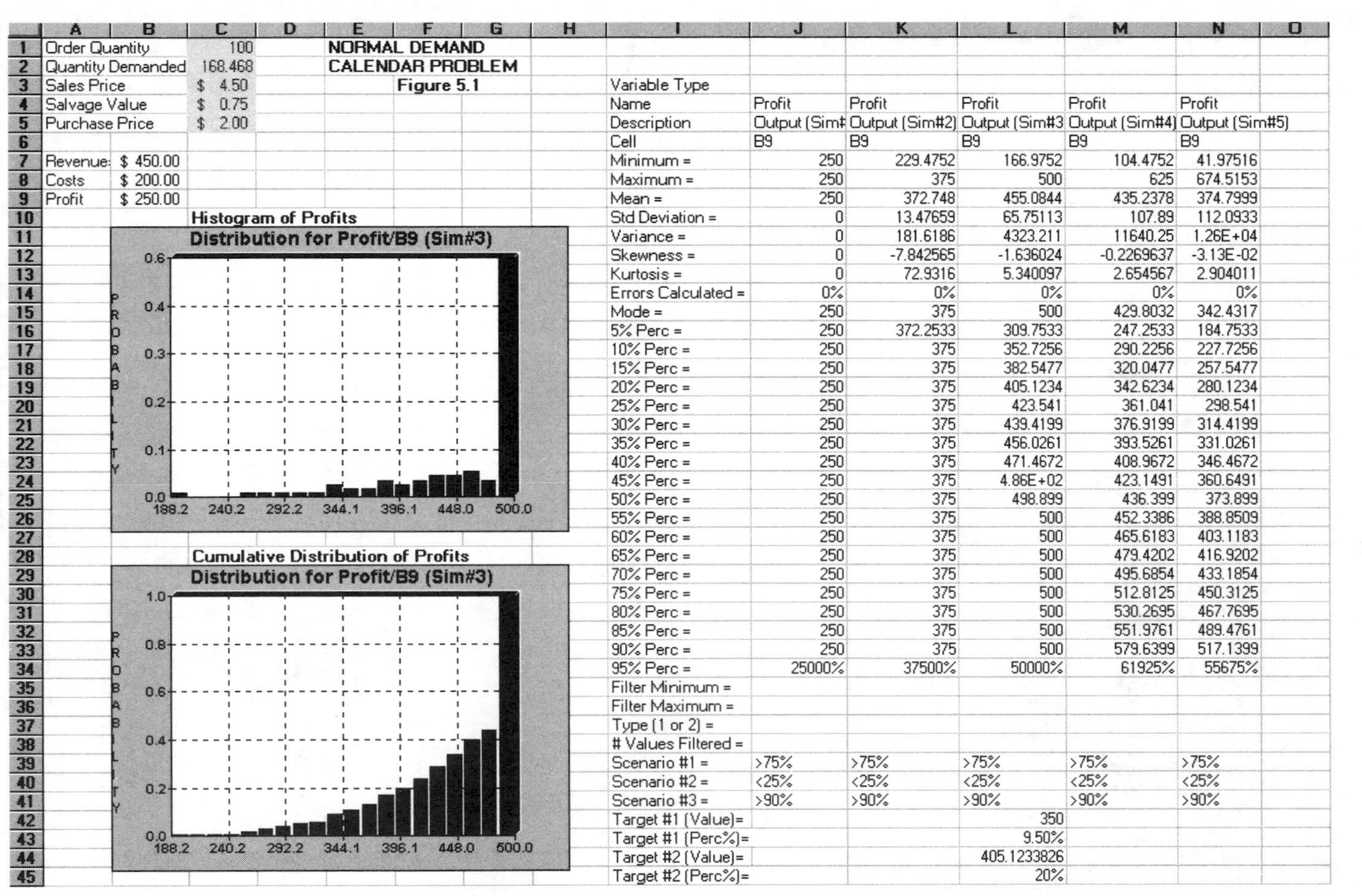

	A	B	C	E–G
1	Order Quantity		100	NORMAL DEMAND
2	Quantity Demanded		168.468	CALENDAR PROBLEM
3	Sales Price		$ 4.50	Figure 5.1
4	Salvage Value		$ 0.75	
5	Purchase Price		$ 2.00	
7	Revenue:	$ 450.00		
8	Costs	$ 200.00		
9	Profit	$ 250.00		

	I	J	K	L	M	N
3	Variable Type					
4	Name	Profit	Profit	Profit	Profit	Profit
5	Description	Output (Sim#	Output (Sim#2)	Output (Sim#3	Output (Sim#4)	Output (Sim#5)
6	Cell	B9	B9	B9	B9	B9
7	Minimum =	250	229.4752	166.9752	104.4752	41.97516
8	Maximum =	250	375	500	625	674.5153
9	Mean =	250	372.748	455.0844	435.2378	374.7999
10	Std Deviation =	0	13.47659	65.75113	107.89	112.0933
11	Variance =	0	181.6186	4323.211	11640.25	1.26E+04
12	Skewness =	0	-7.842565	-1.636024	-0.2269637	-3.13E-02
13	Kurtosis =	0	72.9316	5.340097	2.654567	2.904011
14	Errors Calculated =	0%	0%	0%	0%	0%
15	Mode =	250	375	500	429.8032	342.4317
16	5% Perc =	250	372.2533	309.7533	247.2533	184.7533
17	10% Perc =	250	375	352.7256	290.2256	227.7256
18	15% Perc =	250	375	382.5477	320.0477	257.5477
19	20% Perc =	250	375	405.1234	342.6234	280.1234
20	25% Perc =	250	375	423.541	361.041	298.541
21	30% Perc =	250	375	439.4199	376.9199	314.4199
22	35% Perc =	250	375	456.0261	393.5261	331.0261
23	40% Perc =	250	375	471.4672	408.9672	346.4672
24	45% Perc =	250	375	4.86E+02	423.1491	360.6491
25	50% Perc =	250	375	498.899	436.399	373.899
26	55% Perc =	250	375	500	452.3386	388.8509
27	60% Perc =	250	375	500	465.6183	403.1183
28	65% Perc =	250	375	500	479.4202	416.9202
29	70% Perc =	250	375	500	495.6854	433.1854
30	75% Perc =	250	375	500	512.8125	450.3125
31	80% Perc =	250	375	500	530.2695	467.7695
32	85% Perc =	250	375	500	551.9761	489.4761
33	90% Perc =	250	375	500	579.6399	517.1399
34	95% Perc =	25000%	37500%	50000%	61925%	55675%
35	Filter Minimum =					
36	Filter Maximum =					
37	Type (1 or 2) =					
38	# Values Filtered =					
39	Scenario #1 =	>75%	>75%	>75%	>75%	>75%
40	Scenario #2 =	<25%	<25%	<25%	<25%	<25%
41	Scenario #3 =	>90%	>90%	>90%	>90%	>90%
42	Target #1 (Value)=			350		
43	Target #1 (Perc%)=			9.50%		
44	Target #2 (Value)=			405.1233826		
45	Target #2 (Perc%)=			20%		

Figure 5.1 Calendar Problem with Normal Demand

	A	B	C	D	E	F	G
1	Order Quantity		100		NORMAL DEMAND		
2	Quantity Demanded		168.4678		CALENDAR PROBLEM		
3	Sales Price		$ 4.50			Figure 5.1	
4	Salvage Value		$ 0.75				
5	Purchase Price		$ 2.00				
6							
7	Revenues	$ 450.00					
8	Costs	$ 200.00					
9	Profit	$ 250.00					

Figure 5.1a

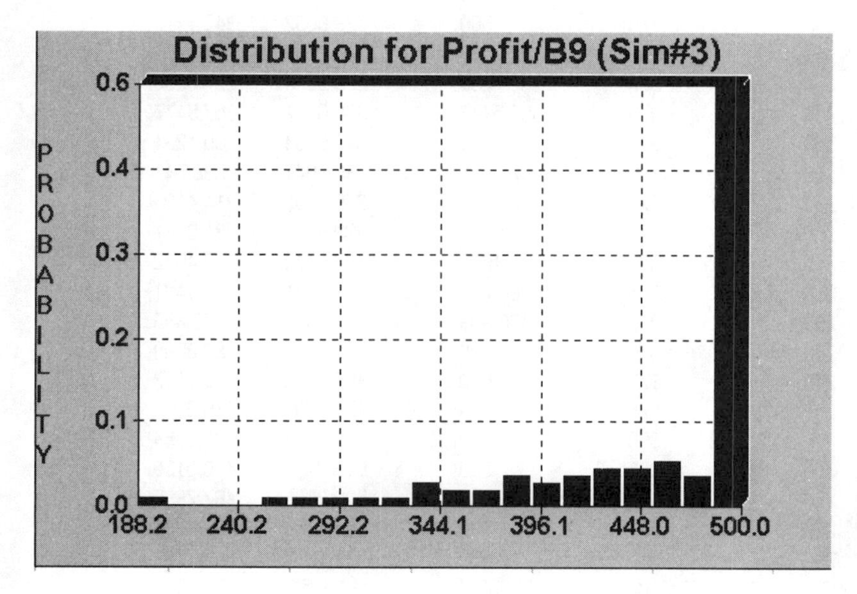

Figure 5.1b

is, @RISK will generate a demand from a normal distribution with mean 200 and standard deviation 30 and ignore it if the demand is less than 0 or more than 1000 (we chose 1000 because there is virtually no chance that demand will exceed 1000).

As in Chapter 4, we now select cell B9 as an output cell. We are still considering five possible order quantities (100, 150, 200, 250, and 300), so we run five simulations. The results in Figure 5.1c indicate that for the five order quantities considered, ordering 200 calendars maximizes expected profit. Since demand can now assume any value, however, we do not know that ordering 200 calendars will maximize expected profit. The fact that

Expected Profit for 150 calendars < Expected Profit for 200 calendars

and

Expected Profit for 200 Calendars > Expected profit for 250 calendars

Variable Type					
Name	Profit	Profit	Profit	Profit	Profit
Description	Output (Sim#1	Output (Sim#2)	Output (Sim#3)	Output (Sim#4)	Output (Sim#5)
Cell	B9	B9	B9	B9	B9
Minimum =	250	229.4752	166.9752	104.4752	41.97516
Maximum =	250	375	500	625	674.5153
Mean =	250	372.748	455.0844	435.2378	374.7999
Std Deviation =	0	13.47659	65.75113	107.89	112.0933
Variance =	0	181.6186	4323.211	11640.25	1.26E+04
Skewness =	0	-7.842565	-1.636024	-0.2269637	-3.13E-02
Kurtosis =	0	72.9316	5.340097	2.654567	2.904011
Errors Calculated =	0%	0%	0%	0%	0%
Mode =	250	375	500	429.8032	342.4317
5% Perc =	250	372.2533	309.7533	247.2533	184.7533
10% Perc =	250	375	352.7256	290.2256	227.7256
15% Perc =	250	375	382.5477	320.0477	257.5477
20% Perc =	250	375	405.1234	342.6234	280.1234
25% Perc =	250	375	423.541	361.041	298.541
30% Perc =	250	375	439.4199	376.9199	314.4199
35% Perc =	250	375	456.0261	393.5261	331.0261
40% Perc =	250	375	471.4672	408.9672	346.4672
45% Perc =	250	375	4.86E+02	423.1491	360.6491
50% Perc =	250	375	498.899	436.399	373.899
55% Perc =	250	375	500	452.3386	388.8509
60% Perc =	250	375	500	465.6183	403.1183
65% Perc =	250	375	500	479.4202	416.9202
70% Perc =	250	375	500	495.6854	433.1854
75% Perc =	250	375	500	512.8125	450.3125
80% Perc =	250	375	500	530.2695	467.7695
85% Perc =	250	375	500	551.9761	489.4761
90% Perc =	250	375	500	579.6399	517.1399
95% Perc =	25000%	37500%	50000%	61925%	55675%
Filter Minimum =					
Filter Maximum =					
Type (1 or 2) =					
# Values Filtered =					
Scenario #1 =	>75%	>75%	>75%	>75%	>75%
Scenario #2 =	<25%	<25%	<25%	<25%	<25%
Scenario #3 =	>90%	>90%	>90%	>90%	>90%
Target #1 (Value)=			350		
Target #1 (Perc%)=			9.50%		
Target #2 (Value)=			405.1233826		
Target #2 (Perc%)=			20%		

Figure 5.1c

indicates that expected profit is maximized by ordering a quantity between 150 and 250 calendars. We could now try another `SIMTABLE` in cell C1 such as

```
Excel: =RISKSIMTABLE({160,170,180,190,210,220,230,240})

Lotus: @RISKSIMTABLE(160,170,180,190,210,220,230,240)
```

After running eight simulations we would then find that expected profit is maximized for an order quantity between 200 and 220 calendars. Actually, it can be analytically shown that expected profit is maximized by ordering 213 calendars.

5.2 Using the Graph Type Command

After obtaining a histogram of possible profits for an order quantity (analagous to Figure 4.1) we may use the @RISK Type command to generate the cumulative probability graph in Figure 5.1a. After obtaining a histogram for profit for Simulation #3, click on the left mouse button to activate the Graph submenu. Now select Type, Cumulative Ascending and Bar Graph to generate the cumulative distribution of profit in Figure 5.1d. For any particular value of profit (read on the x-axis), the y-axis gives the probability that actual profit will be less than or equal to the x-value. For example, there appears to be about a 20% probability that profit will be less than or equal to $405. This can be verified by looking at the percentiles from the statistics section of the results. If you choose a Cumulative Descending graph, the y-coordinate will be the probability that actual profit will be greater than or equal to the x-coordinate.

Figure 5.1d

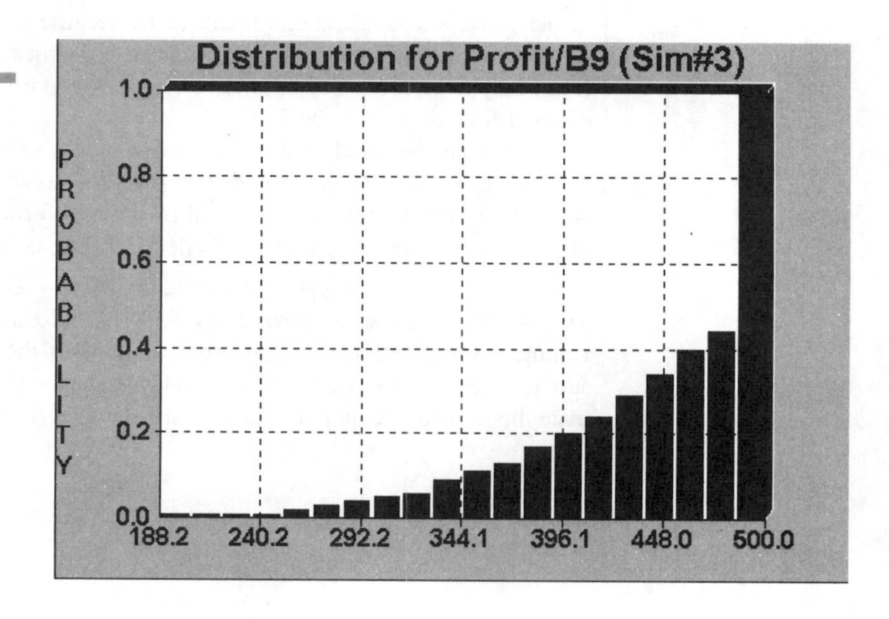

5.3 Placing Target Values in the Statistics Output

For any possible value of profit, @RISK can determine (in the target section of the statistics results) the probability that actual profit will less than or equal to a given "target value." To use the Target option, move down to the bottom of the statistics report and enter any value as a target. @RISK will determine the probability that profit will be less than or equal to the target. For example (for an order quantity of 200), if we enter a target of $350, @RISK tells us that there is a 9.5% chance that our profit will be $350 or less. Conversely, if we enter a probability (as a percentage), then @RISK tells us which profit value has that probability of not being exceeded. For example, if we enter a probability of 20%, @RISK responds with a value of $405.12. This response indicates that there is a 20% chance that profit will be less than or equal to $405.12.

5.4 Estimating the Mean and Standard Deviation of a Normal Distribution

Suppose you feel that annual sales of calendars are symmetric. Then modeling annual sales of calendars by a normal distribution might be appropriate. If you have no data for future sales, how can you subjectively estimate the mean and standard deviation of a normal distribution? To estimate the mean of a normal distribution, pick a number (call it Mean) such that next year's demand for calendar sales is equally likely to be below Mean as above Mean. The response to this request is a good estimate for next year's mean demand for calendars, because it represents the 50th percentile of the normal distribution. Thus, if we believe that next year's demand for calendars is as likely to be below 200 as above 200, we would estimate the mean of next year's demand for calendars to be 200.

To estimate the standard deviation for next year's demand for calendars, find two numbers Lower and Upper (equidistant from Mean) such that you are 95% sure that next year's demand for calendars will be between Lower and Upper. Then estimate the standard deviation of next year's demand for calendars by

$$(\text{Upper} - \text{Mean})/2 \text{ or } (\text{Mean} - \text{Lower})/2$$

This follows because approximately 95% of the time a normal distribution will assume a value within two standard deviations of the mean. Thus, if we are 95% sure that next year between 140 and 260 calendars will be demanded, we would estimate the standard deviation of next year's demand for calendars to be (200 – 140)/2 = 30.

Problems

Group A

5.1 For Problem 3.2 suppose that the demand for cars is normally distributed with $\mu = 40$ and $\sigma = 7$. Use simulation to determine an optimal order quantity. For your optimal order quantity, determine a 95% confidence interval for expected profit.

5.2 Six months before its annual convention, the American Medical Association must determine how many rooms to reserve. At this time the AMA can reserve rooms at a cost of $50 per room. The AMA must pay the $50 room cost even if the room is not occupied. The AMA believes the number of doctors attending the convention will be normally distributed with a mean of 5000 and a standard deviation of 1000. If the number of people attending the convention exceeds the number of rooms reserved, extra rooms must be reserved at a cost of $80 per room. Use simulation to determine the number of rooms that should be reserved to minimize the expected cost to the AMA.

5.3 A ticket from Indianapolis to Orlando on Deleast Airlines sells for $150. The plane can hold 100 people. It costs $8000 to fly an empty plane. Each person on the plane incurs variable costs of $30 (food and fuel). If the flight is overbooked, anyone who cannot get a seat receives $300 in compensation. On the average, 95% of all people who have a reservation show up for the flight. To maximize expected profit, how many reservations for the flight should be taken by Deleast?

Hint: The @RISK function `=RISKBINOMIAL` (or `@RISKBINOMIAL`) can be used to simulate the number of passengers who show up. If the number of reservations taken is in cell A2, then the formula

```
Excel:  =RISKBINOMIAL(A2,.95)
Lotus:  @RISKBINOMIAL(A2,.95)
```

will generate the number of customers who actually show up for a flight!

CHAPTER 6

Applications of Simulation to Corporate Financial Planning

As mentioned in Chapter 1, many companies use simulation in their capital budgeting and financial planning processes. Simulation can be used to model the uncertainty associated with future cash flows and to answer questions such as the following:

- What is the estimated mean and variance of a project's Net Present Value (NPV)?
- What is the estimated probability that a project will have a negative NPV?
- What are the estimated mean and variance of a company's profit during the next fiscal year?
- What is the estimated probability that during the next year a company will have to borrow more than $2 million?

The following example illustrates how simulation can be used to compare investment opportunities.

Example 6.1

General Ford (GF) Auto Corporation is trying to determine what type of compact car to develop. Two models (Model 1 and Model 2) are under consideration. Each model is assumed to generate sales for ten years. In order to determine which model should be built, information about the following quantities has been gathered through focus groups with the marketing and engineering departments.

Fixed Cost of Developing Car: This cost is assumed to be incurred at the beginning of Year 1 (or end of Year 0) before any sales are recorded.

Variable Production Cost: The variable cost incurred in producing a car.

Sales Price: The sales price is assumed to be $10,000 for each model.

Sales of Car During Each of Next Ten Years: For simplicity we will assume that all sales occur at the end of each year.

Interest Rate: It is assumed that cash flows are discounted at 10%. This means that a cash outflow of $1 at the beginning of Year 1 is equivalent to a cash outflow of $1.10 at the end of Year 1.

Fixed and variable costs and annual sales are not known with certainty. The views of marketing and engineering about these quantities are summarized in Table 6.1.

Table 6.1

Fixed Cost for Car 1		**Fixed Cost for Car 2**	
Probability	***Value***	***Probability***	***Value***
.50	$6 billion	.25	$4 billion
.50	$8 billion	.50	$5 billion
		.25	$16 billion
Variable Cost for Car 1		**Variable Cost for Car 2**	
Probability	***Value***	***Probability***	***Value***
.50	$4600	.50	$2000
.50	$5400	.50	$6000
Year 1 Unit Sales for Car 1 (Number of Cars)		**Year 1 Unit Sales for Car 2 (Number of Cars)**	
Probability	***Value***	***Probability***	***Value***
.25	230,000	.25	80,000
.50	250,000	.50	220,000
.25	270,000	.25	390,000

Of course, if a car sells well the first year, it probably will sell well during later years. GF models this belief by assuming that

$$\text{Expected Year } t \text{ Sales for a Model} = \text{Actual Year } t - 1 \text{ Sales} \qquad \textbf{(6.1)}$$

Then GF assumes that

$$\text{Actual Year } t \text{ Sales for a Model} = \text{Expected Year } t \text{ Sales} + \text{Error Term}$$

For Model 1 we assume that the error term is normally distributed with a mean of 0 and a standard deviation of 20,000. For Model 2 we assume that the error term is normally distributed with a mean of 0 and a standard deviation of 30,000. The **error term** models the variability of each model's sales about the expected sales for each year. This error term might be estimated by seeing how well (6.1) has modeled sales during past years. For example, suppose in the past we have predicted next year's sales to equal last year's sales. If these predictions yield a standard deviation for forecast errors of 30,000 cars per year, we can use this estimate in our simulation.

For simplicity, we assume that the variable cost for each year's production is the same. This ignores inflation (see Problem 6.2). How do we use simulation to compare the merits of the two proposed models?

Solution Figure 6.1 (file Npv1.wk4 or Npv1.xls) contains the spreadsheet used to simulate the NPV for Model 1; Figure 6.3 (file Npv2.wk4 or Npv2.xls) contains the spreadsheet used to simulate the NPV for Model 2. We will describe the development of the spreadsheet for Model 1 and leave Model 2 to the reader (see Problem 6.1).

We begin by entering the inputs for our simulation in the cell range B9:B14.

Step 1 In cell B9 of Figure 6.1a we generate Year 1 sales by entering in cell B9 the formula

```
Excel: =RISKDISCRETE({230000,250000,270000},{.25,.5,.25})
Lotus: @RISKDISCRETE(230000,,250000,,270000,.25,.5,.25)
```

Step 2 In cell B11 we generate the variable cost for each car by entering the statement

```
Excel: =RISKDISCRETE({4600,5400},{.5,.5})
Lotus: @RISKDISCRETE(4600,5400,.5,.5)
```

This ensures that there is an equal chance that the variable cost of each car will be $4600 or $5400.

Step 3 In B10 we generate the fixed cost associated with Model 1 by entering the statement

```
Excel: =RISKDISCRETE({6000000000,8000000000},{.5,.5})
Lotus: @RISKDISCRETE(6000000000,8000000000,.5,.5)
```

This ensures that fixed cost is equally likely to be $6 or $8 billion.

Step 4 In cell B12 we enter the sales price ($10,000) for each car. Then in cell B13 we enter the interest rate (.1 or 10%). Finally, in cell B14 we enter the standard deviation (20,000 cars) of annual unit sales about last year's actual sales.

Step 5 In cell C3 we recopy Year 1 sales by entering

```
Excel: =B9
Lotus: +B9
```

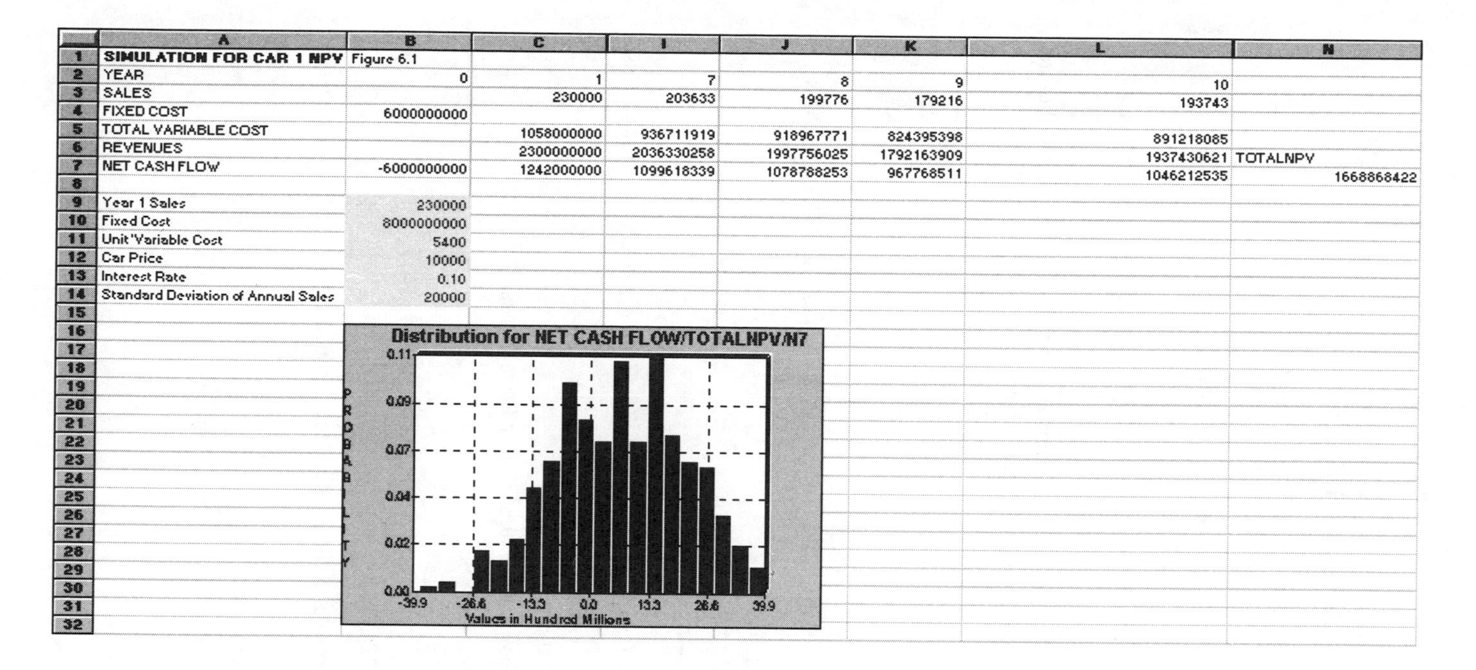

	A	B	C	I	J	K	L	N
1	**SIMULATION FOR CAR 1 NPV**	Figure 6.1						
2	YEAR	0	1	7	8	9	10	
3	SALES		230000	203633	199776	179216	193743	
4	FIXED COST	6000000000						
5	TOTAL VARIABLE COST		1058000000	936711919	918967771	824395398	891218085	
6	REVENUES		2300000000	2036330258	1997756025	1792163909	1937430621	TOTALNPV
7	NET CASH FLOW	-6000000000	1242000000	1099618339	1078788253	967768511	1046212535	1668868422
8								
9	Year 1 Sales	230000						
10	Fixed Cost	8000000000						
11	Unit 'Variable Cost	5400						
12	Car Price	10000						
13	Interest Rate	0.10						
14	Standard Deviation of Annual Sales	20000						
15								
16								
17								
18								
19								
20								
21								
22								
23								
24								
25								
26								
27								
28								
29								
30								
31								
32								

Figure 6.1 Car 1 Simulation of NPV

	A	B	C	I	J	K	L	N
1	**SIMULATION FOR CAR 1 NPV**	Figure 6.1						
2	YEAR	0	1	7	8	9	10	
3	SALES		230000	203633	199776	179216	193743	
4	FIXED COST	6000000000						
5	TOTAL VARIABLE COST		1058000000	936711919	918967771	824395398	891218085	
6	REVENUES		2300000000	2036330258	1997756025	1792163909	1937430621	TOTALNPV
7	NET CASH FLOW	-6000000000	1242000000	1099618339	1078788253	967768511	1046212535	1668868422
8								
9	Year 1 Sales	230000						
10	Fixed Cost	8000000000						
11	Unit 'Variable Cost	5400						
12	Car Price	10000						
13	Interest Rate	0.10						
14	Standard Deviation of Annual Sales	20000						

Figure 6.1a

Step 6 Recall that

Year 2 Sales = Year 1 Sales + (Normal Random Variable with Mean 0 and σ = 20,000)

To model this relationship we generate Year 2 sales (in cars sold, not dollar value) by entering into cell D3 the formula

```
Excel:  =C3+RISKNORMAL(0,$B14)(or
        =RISKNORMAL(C3,$B14))

Lotus:  +C3+@RISKNORMAL(0,$B14)(or
        @RISKNORMAL(C3,$B14))
```

Copying this formula from cell D3 to E3:L3 generates sales for Years 3–10.

Step 7 In B4 we reenter the fixed cost (assumed to be incurred in Year 0, before the sales of any cars occur) with the formula

```
Excel:  =B10

Lotus:  +B10
```

Step 8 In C5 we determine

Year 1 Total Variable Cost = (Variable Cost per Car)*(Year 1 Sales)

by entering

```
Excel:  =C3*$B11

Lotus:  +C3*$B11
```

Copying this formula to the range D5:L5 generates the total variable cost incurred during each year.

Step 9 In cell C6 we compute

Year 1 Revenues = (Car Price)*(Year 1 Sales)

by entering the formula

```
Excel:  =$B12*C3

Lotus:  +$B12*C3
```

Copying this formula to the range D6:L6 generates sales revenue received during years 2–10.

Step 10 In cell B7 we compute

Year 0 Net Cash Flow = – (Fixed Cost)

by entering

```
Excel:  =-B4

Lotus:  -B4
```

Step 11 In cell C7 we generate

Year 1 Net Cash Flow = Year 1 Revenues – Year 1 Variable Costs

by entering

```
Excel: =C6-C5

Lotus: +C6-C5
```

Copying this formula to the range D7:L7 generates net cash flow for Years 2–10.

Step 12 In cell N7 we determine the NPV of all cash flows by entering the statement

```
Excel: =B7+NPV(B13,C7:L7)(or
       =(1+B13)*NPV(B13,B7:L7))

Lotus: +B7+@NPV(B13,C7..L7)(or
       (1+B13)*@NPV(B13,B7..L7))
```

This computes

$$\text{Net Cash Flow Year 0} + \frac{\text{Net Cash Flow Year 1}}{(1+0.1)} + \frac{\text{Net Cash Flow Year 2}}{(1+0.1)^2} + \ldots \frac{\text{Net Cash Flow Year 10}}{(1+0.1)^{10}}$$

which is the NPV of all cash flows. Note that the first argument in the `NPV(  )` function is the relevant interest rate, and the second argument is the range containing the cash flows received during each time period.

Step 13 Select cell N7 as our output cell and run 400 iterations with @RISK. See Figure 6.2 for the results of the simulation. A histogram of the NPVs for the 400 iterations is given in Figure 6.1b. After running 400 iterations of Model 2 we obtained the results in Figure 6.3. Note that to simulate Model 2 we need only change our inputs for Model 1 in cells B9, B10, B11, and B14. Table 6.2 summarizes our results. (We used the `TARGET` command to obtain the probabilities for the NPV of each model.)

Figure 6.1b

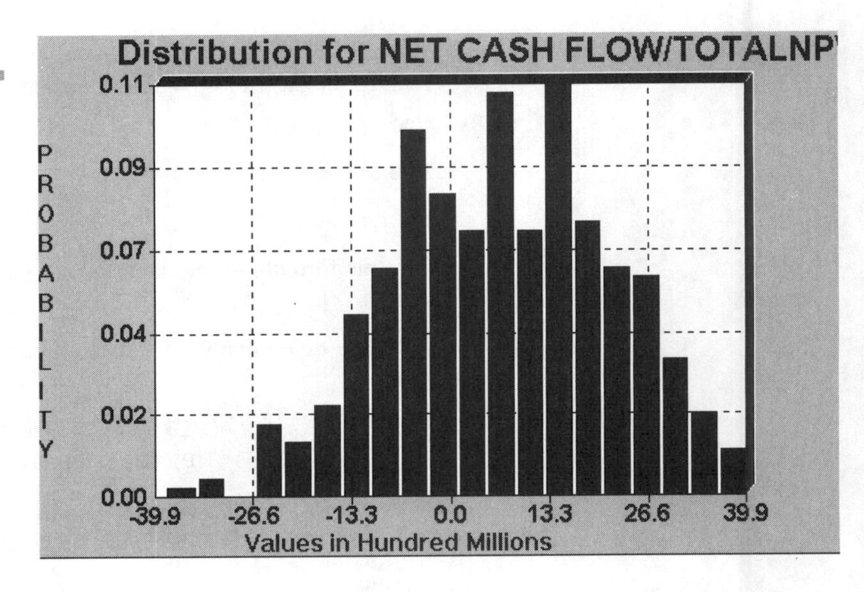

Table 6.2 Results for Car Simulation

	Model 1	**Model 2**
Average NPV	$682 million	$990 million
Standard Deviation	1.47 billion	7.40 billion
Probability NPV < 0	34.25%	43.38%
Probability NPV < –1 billion	12.8%	35.67%

The simulation indicates that Model 1 has a lower expected NPV than Model 2. Model 1 also has a much lower standard deviation than Model 2. This indicates that on the average, Model 2 yields a higher NPV but is also a much riskier proposition than Model 1. This is confirmed by the fact that for our simulation, Model 1 has only an estimated 12.8% chance of losing more than $1 billion (in NPV terms), while Model 2 has an estimated 35.67% chance of losing more than $1 billion. If your company is strongly risk averse (see Section 12.2 of Winston 1994), then Model 1 is the way to go.

Remark

In Step 2 the variable cost for each car could also have been modeled by the statement

```
Excel: =RISKDUNIFORM({4600,5400})

Lotus: @RISKDUNIFORM(4600,5400)
```

The `RISKDUNIFORM` function assigns equal probability to each of its arguments.

	B	C
2	**Figure 6.2**	
3	Variable Type	
4	Name	NPV
5	Description	Output
6	Cell	N7
7	Minimum =	-3,818,588,000.00
8	Maximum =	3,985,339,000.00
9	Mean =	681,828,900.00
10	Std Deviation =	1,470,573,000.00
11	Variance =	2,162,585,000,000,000,000.00
12	Skewness =	-0.13
13	Kurtosis =	2.55
14	Errors Calculated =	0.00
15	Percentile Values	
16	5% Perc =	-1,698,810,000.00
17	10% Perc =	-1,216,098,000.00
18	15% Perc =	-855,803,100.00
19	20% Perc =	-619,830,800.00
20	25% Perc =	-428,018,000.00
21	30% Perc =	-260,561,800.00
22	35% Perc =	35,884,580.00
23	40% Perc =	296,303,800.00
24	45% Perc =	533,823,700.00
25	50% Perc =	655,542,500.00
26	55% Perc =	881,750,300.00
27	60% Perc =	1,131,694,000.00
28	65% Perc =	1,363,050,000.00
29	70% Perc =	1,526,492,000.00
30	75% Perc =	1,748,060,000.00
31	80% Perc =	2,009,885,000.00
32	85% Perc =	2,316,867,000.00
33	90% Perc =	2,568,141,000.00
34	95% Perc =	3,077,001,000.00
35	Filter Minimum =	
36	Filter Maximum =	
37	Type (1 or 2) =	
38	# Values Filtered =	
39	Scenario #1 =	>75%
40	Scenario #2 =	<25%
41	Scenario #3 =	>90%

Figure 6.2 Statistics for Car 1 Simulation

	A	B	C	I	J	K	L	N
1	**SIMULATION FOR CAR 2 NPV**		**Figure 6.3**					
2	YEAR	0	1	7	8	9	10	
3	SALES		80000	168006	197724	175858	194055	
4	FIXED COST	5000000000						
5	TOTAL VARIABLE COST		480000000	1008033763	1186346164	1055147665	1164330433	
6	REVENUES		800000000	1680056272	1977243606	1758579442	1940550721	TOTALNPV
7	NET CASH FLOW	-5000000000	320000000	672022509	790897443	703431777	776220288	-2057763619
8								
9	Year 1 Sales	80000						
10	Fixed Cost	5000000000						
11	Unit 'Variable Cost	6000						
12	Car Price	10000						
13	Interest Rate	0.1						
14	Standard Deviation of Annual Sale	30000						
15								
16		Variable Type						
17		Name	NET CASH FLOW/TOTALNPV					
18		Description	Output					
19		Cell	N7					
20		Minimum =	-16,872,350,000.00					
21		Maximum =	21,164,810,000.00					
22		Mean =	989,990,300.00					
23		Std Deviation =	7,399,980,000.00					
24		Variance =	54,759,690,000,000,000,000.00					
25		Skewness =	0.17					
26		Kurtosis =	2.91					
27		Errors Calculated =	0.00					
28		Percentile Values						
29		5% Perc =	-11,915,390,000.00					
30		10% Perc =	-9,815,012,000.00					
31		15% Perc =	-6,756,889,000.00					
32		20% Perc =	-4,919,934,000.00					
33		25% Perc =	-3,577,965,000.00					
34		30% Perc =	-2,106,874,000.00					
35		35% Perc =	-1,051,135,000.00					
36		40% Perc =	-450,571,400.00					
37		45% Perc =	167,144,800.00					
38		50% Perc =	807,940,000.00					
39		55% Perc =	1,265,608,000.00					
40		60% Perc =	2,090,490,000.00					
41		65% Perc =	3,372,805,000.00					
42		70% Perc =	4,367,617,000.00					
43		75% Perc =	5,163,311,000.00					
44		80% Perc =	5,888,766,000.00					
45		85% Perc =	7,316,252,000.00					
46		90% Perc =	12,650,340,000.00					
47		95% Perc =	15,089,010,000.00					
48		Filter Minimum =						
49		Filter Maximum =						

Figure 6.3 Simulation of NPV for Car 2

6.1 Using the Triangular Distribution to Model Sales

It is unrealistic to assume that Year 1 sales must always equal 230,000, 250,000, or 270,000 cars. A more realistic model might allow car sales for Year 1 to assume any integer value between 230,000 and 270,000. To model this possibility we enter into cell B9 the statement

```
Excel:  =RISKTRIANG(230000,250000,270000)

Lotus:  @RISKTRIANG(230000,250000,270000)
```

This ensures that Year 1 sales are drawn from the distribution pictured in Figure 6.4. For obvious reasons this is called a **triangular distribution**. The first input to the `@TRIANG` statement is the smallest value (most pessimistic outcome) the random variable can assume; the second input is the most likely value the random variable can assume; and the final input is the largest value (most optimistic outcome) the random variable can assume. It is useful to model a random variable by a triangular distribution when you have a good guess at the best, worst, and most likely scenarios.

To get an idea of the probabilities implied by a triangular distribution, recall that for any continuous random variable the probability that the random variable assumes a value between two numbers a and b is the area under the density function between

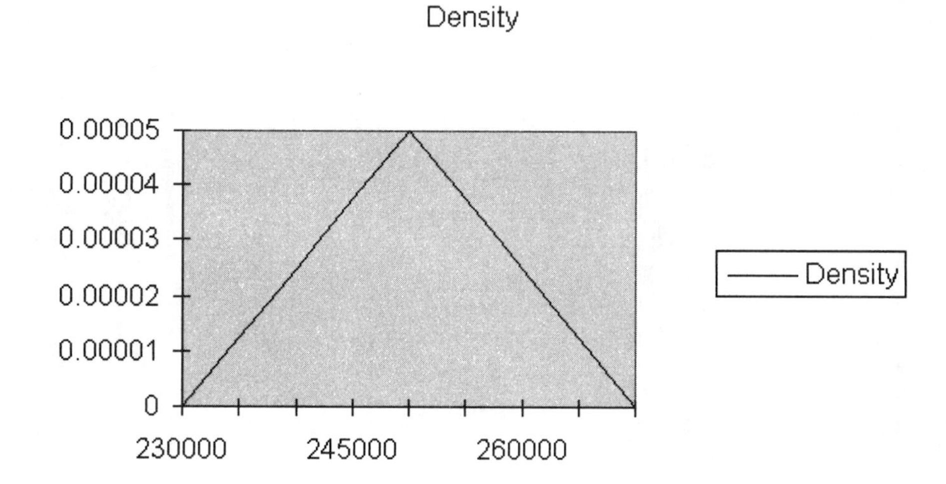

Figure 6.4 Triangular Distribution

	A	B	C	D	E	F	G	H	I	J	K	L	M	N
1	**Simulation for Car 1 NPV;triangular distribution**													
2	YEAR	0	1	2	3	4	5	6	7	8	9	10		
3	SALES		267968	287086	270893	253836	241553	217552	194627	187326	167770	144160		
4	FIXED COST	8000000000												
5	TOTAL VARIABLE COST		1232652672	1.3E+09	1.2E+09	1.2E+09	1.1E+09	1E+09	9E+08	8.6E+08	7.7E+08	6.6E+08		
6	REVENUES		2679679722	2.9E+09	2.7E+09	2.5E+09	2.4E+09	2.2E+09	1.9E+09	1.9E+09	1.7E+09	1.4E+09	TOTALNPV	
7	NET CASH FLOW	-8E+09	1447027050	1.6E+09	1.5E+09	1.4E+09	1.3E+09	1.2E+09	1.1E+09	1E+09	9.1E+08	7.8E+08	-2E+08	
8														
9	Year 1 Sales	267968												
10	Fixed Cost	8000000000												
11	Unit 'Variable Cost	4600												
12	Car Price	10000												
13	Interest Rate	0.10												
14	Standard Deviation of Annual Sales	20000												
15														
16		Variable Type												
17		Name	NET CASH FLOW/TOTALNPV											
18		Description	Output											
19		Cell	N7											
20		Minimum =	-3,355,664,000.00											
21		Maximum =	4,775,675,000.00											
22		Mean =	681,989,900.00											
23		Std Deviation =	1,568,605,000.00											
24		Variance =	2,460,520,000,000,000,000.00											
25		Skewness =	0.16											
26		Kurtosis =	2.52											
27		Errors Calculat	0											
28		Percentile Values												
29		5% Perc =	-1,798,674,000.00											
30		10% Perc =	-1,239,359,000.00											
31		15% Perc =	-1,030,104,000.00											
32		20% Perc =	-756,387,500.00											
33		25% Perc =	-513,919,800.00											
34		30% Perc =	-263,578,000.00											
35		35% Perc =	-50,861,160.00											
36		40% Perc =	135,381,400.00											
37		45% Perc =	338,161,900.00											
38		50% Perc =	614,449,700.00											
39		55% Perc =	869,309,800.00											
40		60% Perc =	1,074,888,000.00											
41		65% Perc =	1,399,712,000.00											
42		70% Perc =	1,582,274,000.00											
43		75% Perc =	1,790,236,000.00											
44		80% Perc =	2,007,819,000.00											
45		85% Perc =	2,386,433,000.00											
46		90% Perc =	2,799,207,000.00											
47		95% Perc =	3,358,348,000.00											
48		Filter Minimum =												
49		Filter Maximum =												

Figure 6.5 Car 1 Simulation with Year 1 Sales Triangular

a and *b*. For example, Figure 6.4 implies that the probability that Year 1 sales total at most 240,000 cars equals

$$.5*(240{,}000 - 230{,}000)*(.000025) = .125$$

and the probability that Year 2 sales total between 240,000 and 250,000 cars equals

$$.5*(250{,}000 - 240{,}000)*(.00005 + .000025) = .375$$

In Figure 6.5 (file Triang1.wk4 or Triang1.xls) we give the @RISK results for the simulation. Observe that the average NPV and its standard deviation are almost unchanged from the situation in which sales were modeled as a discrete random variable.

In our triangular distribution, the most pessimistic and most optimistic outcomes are the same distance from the most likely outcome. It is unnecessary to make this assumption.

6.2 Sensitivity Analysis with Tornado Graphs

A question of natural interest is which inputs to the simulation have the most effect on our output (NPV). Using the @RISK tornado graph we can answer this question. Clicking from the Results menu on Sensitivity, then clicking on Graph and then selecting Correlation or Regression yields the **tornado graphs** in Figures 6.6 and 6.7. The regression tornado graph is created as follows: Using the results of our 400 iterations as data, @RISK runs a multiple regression using the NPV for each iteration as the dependent variable and using the values for each iteration of the cells that are "random" (fixed cost, variable cost per unit, sales for each year) as the independent variables. Then @RISK graphs the standardized regression (or beta) coefficients for each of the independent variables. A **beta coefficient** for an independent variable indicates the number of standard deviations by which the dependent variable increases if the independent variable increases by one standard deviation (assuming all other independent variables are held constant). Thus, increasing fixed cost by one standard deviation will decrease NPV by .676 standard deviations, increasing variable cost by one standard deviation will decrease NPV by .418 standard deviations, etc. The most influential variables are those with the largest coefficients (in absolute value) in the tornado graph. Thus fixed cost and variable cost per car are the most influential variables on NPV, with Year 2 and Year 3 sales right behind.

The correlation tornado graph plots the correlation between any cell that is "random" and the output cell (NPV). The variables having the most influence on the dependent variable are those with the largest (in absolute value) correlation with the dependent variable. From the correlation tornado graph in Figure 6.7 we find that fixed cost and unit variable cost are again the most influential variables, followed by Year 3 sales and Year 2 sales.

Figure 6.6 Regression Tornado Graph

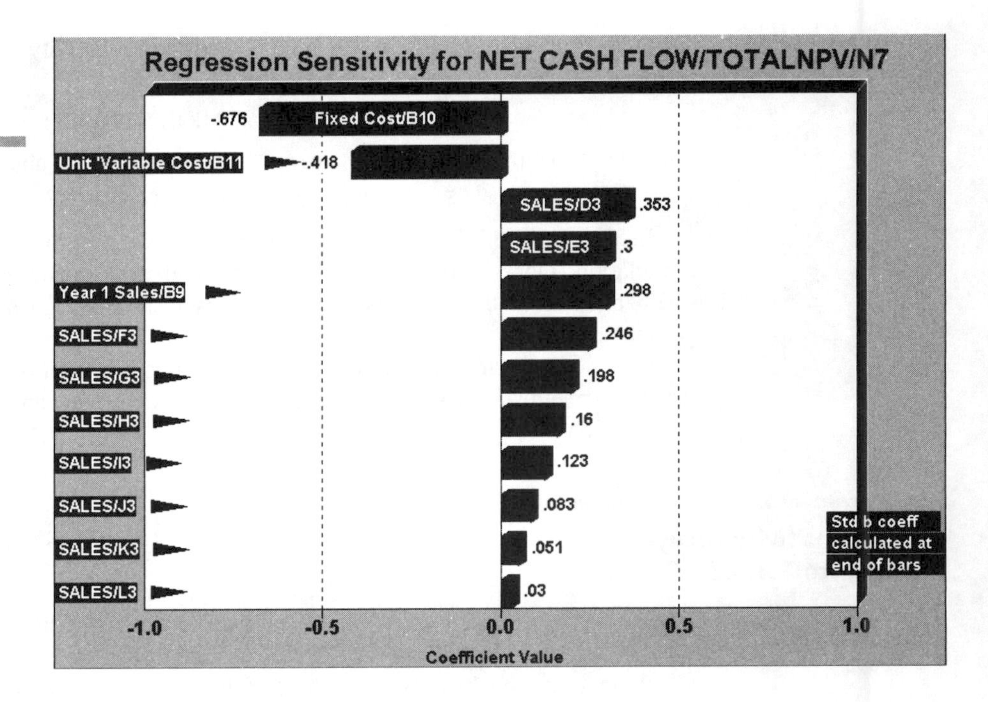

Figure 6.7 Correlation Tornado Graph

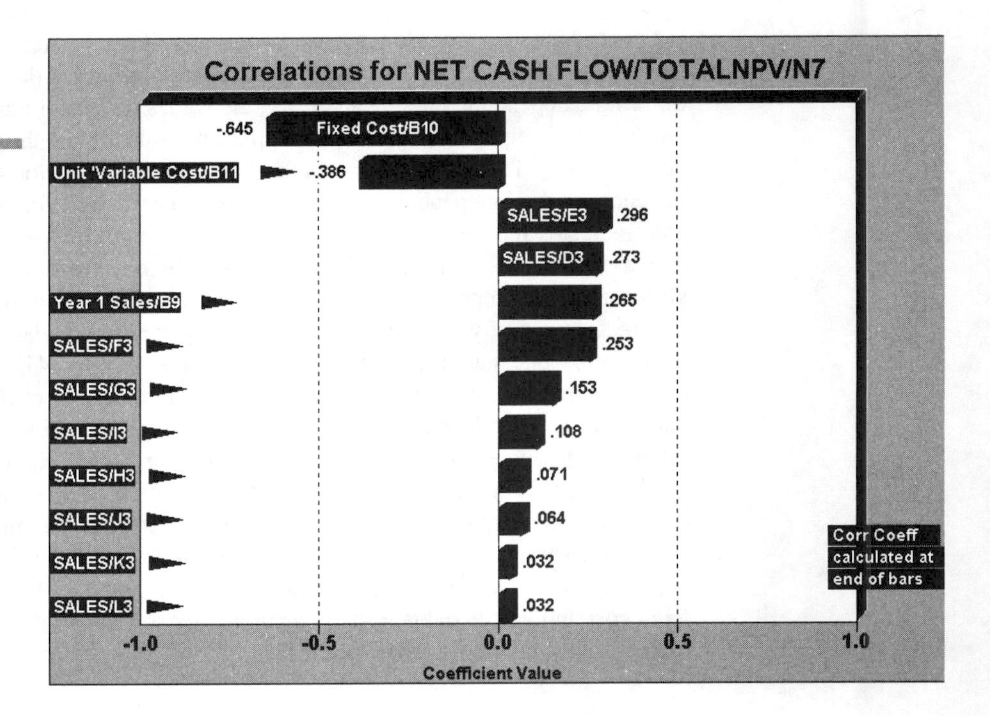

6.3 Sensitivity Analysis with Scenarios

You can also use the @RISK Scenario command to do sensitivity analysis. For each output cell you can enter up to three scenarios. To perform a sensitivity analysis move to the bottom of the statistics results. The default scenarios are the 10th, 75th, and 90th percentile for the output cell. For each scenario @RISK will look at iterations that "meet" the scenario and identify any "random" cell whose median value for iterations meeting the scenario differs by .5 standard deviations or more from its overall median. To access Scenarios click on Scenarios from the Results menu. If you highlight the cell(s) in the top part of the scenario report, you will obtain a more detailed analysis in the bottom half of the screen. @RISK will identify random cells whose values appear to be atypical when a scenario is met. For the NPV > 75% scenario, @RISK identifies cell E3 (Year 3 sales) as being .5 standard deviations above average. This means that for the iterations in which NPV was in the top 25% of all iterations, the median of Year 3 sales were .5 standard deviations above the median for all iterations.

For the NPV > 90%, @RISK identified cell D3 (Year 2 sales) as having a median value of .65 standard deviations above the overall median for iterations in which NPV was in the top 10% of all iterations. @RISK also identified cell E3 (Year 3 sales) as having a median value of .62 standard deviations above the overall median for iterations in which NPV was in the top 10% of all iterations.

For the NPV < 25% scenario @RISK identified cells B11 (Unit Variable Cost) and B10 (Unit Fixed Cost) as both having a median value 2 standard deviations above the overall median for iterations in which NPV was in the bottom 25% of all iterations. Thus we see that the Scenario command allows us to zero in on variables that assume atypical values when a particular scenario occurs. For example, we see that when NPV is very low, both fixed and variable costs tend to be much higher than average. When NPV is very high, sales during Years 2 and 3 tend to be much higher than average.

Remarks

1. If you are going to create a tornado graph or a scenario analysis you must have the Collect Distributions Sample setting on.
2. Unless you run a very large number of iterations, each time you run a simulation on the same worksheet, the tornado graphs and scenario analysis may yield substantially different results.

6.4 Alternative Modeling Strategies

The key to our model was the formulas in row 3 that are used to model the dependence of Year t (unit) sales on Year $t-1$ (unit) sales. Two alternative approaches are the following:

Approach 1: Simply copy the Year 1 Sales formula in B9 to the range D3:L3. This will cause the unit sales for each year to become independent random variables, which is probably inappropriate. It is reasonable to assume that if Year 1 sales are high, Year 2 sales are also likely to be high. This linkage is excluded by Approach 1.

Approach 2: If you feel that sales grow by an average of 5% per year, enter 1.05 in cell B15 and replace the formula in cell D3 with

```
Excel:  =$B15*C3+RISKNORMAL(0,$B14)

Lotus:  +$B15*C3+@RISKNORMAL(0,$B14)
```

Copying this formula to the range E3:L3 will ensure that the average sales grow by 5% per year.

If you are not sure of the duration of car sales, you could generate a random number of years for which the car is sold in cell B16. Then change the formula in cell D7 to

```
Excel:  =IF(D2<=$B16,D5-D6,0)

Lotus:  @IF(D2<=$B16,D5-D6,0)
```

Copying this formula to the cell range E7:L7 ensures that we will not accrue any revenues or costs after sales of the car end.

We can also model the entry and exit of competitors into the industry. For example, we can make the probability that a competitor enters or exits the industry during Year t depend on our Year t profit.

Problems

Group A

6.1 Simulate the NPV for Model 2 400 times.

6.2 Dord Motors is considering whether to introduce a new model: the Racer. The profitability of the Racer will depend on the following factors:

- Fixed Cost of Developing Racer: Equally likely to be \$3 or \$5 billion.

- Sales: Year 1 Sales will be normally distributed with μ = 200,000 and σ = 50,000.

  Year 2 Sales are normally distributed with μ = Year 1 Sales and σ = 50,000.

  Year 3 Sales are normally distributed with μ = Year 2 Sales and σ = 50,000.

  For example, if Year 1 Sales = 180,000, then the Mean for Year 2 Sales will be 180,000.

- Price: Year 1 Price = $13,000

  Year 2 Price = 1.05*{(Year 1 Price) + $30*(% by which Year 1 Sales exceed expected Year 1 Sales)}

  The 1.05 is the result of inflation!

  Year 3 Price = 1.05*{(Year 2 Price) + $30*(% by which Year 2 Sales exceed expected Year 2 Sales)}

  For example, if Year 1 Sales = 180,000, then Year 2 Price = 1.05*{13,000 + 30(– 10)} = $13,335.

- Variable Cost per Car: During Year 1, the variable cost per car is equally likely to be $5000, $6000, $7000, or $8000.

  Variable Cost for Year 2 = 1.05*(Year 1 Variable Cost)

  Variable Cost for Year 3 = 1.05*(Year 2 Variable Cost)

Your goal is to estimate the NPV of the new car during its first three years. Assume cash flows are discounted at 10%; that is, $1 received now is equivalent to $1.10 received a year from now.

a Simulate 400 iterations and estimate the mean and standard deviation of the NPV for the first three years of sales.

I am 95% sure the expected NPV of this project is between ______ and ______.

b Use the Target option to determine a 95% confidence interval for the actual NPV of the Racer during its first three years of operation.

c Use a tornado graph to analyze which factors are most influential in determining the NPV of the Racer.

6.3 Trucko produces the Goatco truck. They want to gain information about the discounted profits earned during the next three years. During a given year the total number of trucks sold in the U.S. is 500,000 + 50,000*GNP – 40,000*INF

where

GNP = % increase in GNP during year

INF = % increase in Consumer Price Index during year

During the next three years Value Line has made the predictions given in Table 6.3.

Table 6.3

Year	1	2	3
GNP	3%	5%	4%
INF	4%	7%	3%

In the past, 95% of Value Line's GNP predictions have been accurate within 6% of the actual GNP increase, and 95% of Value Line's INF predictions have been accurate within 5% of the actual inflation increase. At the beginning of each year, a number of competitors may enter the trucking business. At the beginning of a year the probability that a certain number of competitors will enter the trucking business is given in Table 6.4.

Table 6.4

Number of Competitors	Probability
0	.50
1	.30
2	.10
3	.10

Before competitors join the industry at the beginning of Year 1 there are two competitors. During a year that begins (after competitors have entered the business, but before any have left) with c competitors, Goatco will have a market share given by $.5*(.9)^c$. At the end of each year there is a 20% chance that each competitor will leave the industry. The sales price of the truck and production cost per truck are given in Table 6.5.

Table 6.5

	Year 1	Year 2	Year 3
Sales Price	$15,000	$16,000	$17,000
Variable Cost	$12,000	$13,000	$14,000

a Simulate 500 times the next three years of Truckco's profit. Estimate the mean and variance of the discounted three year profits (use discount rate of 10%).

b Do the same if during each year there is a 50% chance that each competitor leaves the industry.

Hint: You can model the number of firms leaving the industry in a given period with the

```
Excel: =RISKBINOMIAL
Lotus: @RISKBINOMIAL
```

function. For example, if the number of competitors in the industry is in cell A8, then the number of firms leaving the industry during a period can be modeled with the statement `@RISKBINOMIAL(A8,.20)`. Just remember that the `=RISKBINOMIAL` function is not defined if its first argument equals 0.

Group B

6.4 You have the opportunity to buy a project that yields at the end of Years 1–5 the following (random) cash flows:

End of Year 1 Cash Flow is normal with Mean 1000 and Standard Deviation 200.

for t>1, End of Year t Cash Flow is normal with Mean = Actual End of Year $t - 1$ Cash Flow and Standard Deviation = .2 * (Mean of Year t Cash Flow).

a Assuming cash flows are discounted at 10%, determine the expected NPV (in Time 0 dollars) of the cash flows of this project.

b Suppose we are given the following option: at the end of Years 1, 2, 3, or 4 we may give up our right to future cash flows. In return for doing this we receive the following *abandonment value*.

Table 6.6

Time Abandoned	Value Received
End of Year 1	$3000
End of Year 2	$2600
End of Year 3	$1900
End of Year 4	$900

Assume that we make the abandonment decision as follows: we abandon if and only if the expected NPV of the cash flows from the remaining years is smaller than the abandonment value. For example, suppose end of Year 1 cash flow is $900. At this point in time our best guess is that cash flows from Years 2–5 will also be $900. Thus we would abandon the project at the end of Year 1 if $3000 exceeded the NPV of receiving $900 for four straight years. Otherwise we would continue. What is the expected value of the abandonment option?

6.5 Mattel is developing a new Madonna doll. They have made the following assumptions:

It is equally likely that the doll will sell for two, four, six, eight, or ten years.

At the beginning of Year 1 the potential market for the doll is 1 million. The potential market grows by an average of 5% per year. You are 95% sure that the growth in the potential market during any year will be between 3% and 7%.

We believe our "share" of the potential market during Year 1 will be at worst 20%, most likely 40%, and at best 50%. All values between 20% and 50% are possible.

The variable cost of producing a doll during Year 1 is equally likely to be $4 or $6.

The sales price of the doll during Year 1 will be $10.

Each year the sales price and variable cost of producing the doll will increase by 5%.

The fixed cost of developing the doll (incurred in Year 0) is equally likely to be $4, $8, or $12 million.

At Time 0 there is one competitor in the market. During each year that begins with four or fewer competitors there is a 20% chance that a new competitor will enter the market.

To determine Year t unit sales (for $t > 1$) we proceed as follows: Suppose that at the end of Year $t - 1$, x competitors were present. Then we assume that during Year t, a fraction $.9 - .1*x$ of our loyal customers (last year's purchasers) will buy a doll during the next year and a fraction $.2 - .04*x$ of people currently in the market who did not purchase a doll last year will purchase a doll from us this year. We can now generate a prediction for Year t unit sales. Of course, this prediction will not be precise. We assume that it is sure to be accurate within 15%, however.

Cash flows are discounted at 10% per year.

a Estimate the *expected* NPV (in Time 0 dollars) of this project.

You are 95% sure the expected NPV of this project is between _____ and _____.

b You are 95% sure that the *actual* NPV of the project is between _____ and _____.

c What two factors does the tornado diagram indicate are key drivers of the project's profitability?

CHAPTER 7

Simulating a Cash budget

When a firm is planning its cash budget for the next year, it is important to have some idea of how much cash it will need to borrow at different times of the year. The following example shows how @RISK can be used to answer such questions.

Example 7.1 Cash Budgeting

Widgetco believes that sales (in thousands of dollars) for the months November 1997 through July 1998 will each follow a normal distribution with the means and standard deviations given in Table 7.1.

Table 7.1 Monthly Sales for Widgetco

Month	Nov	Dec	Jan	Feb	March	April	May	June	July
Mean	1500	1600	1800	1500	1900	2600	2400	1900	1300
Std. Dev.	70	75	80	80	100	125	120	90	70

Each month Widgetco incurs fixed costs of $250,000. In March taxes of $150,000 and in June taxes of $50,000 must be paid. Dividends of $50,000 are paid in June.

Widgetco estimates that during a given month

(Receipts) = .20*(Current Monthly Sales)
+ .60*(Last Month's Sales) **(7.1)**
+ .20*(Sales from 2 Months Ago)

Materials and labor needed to produce a month's sales must be purchased one month in advance, and the cost of materials and labor averages 80% of the product's sales price. At the beginning of January 1998, Widgetco has $250,000 in cash. Widgetco would like each month's ending cash balance to never dip below $250,000. Thus, for example, during a month in which Widgetco's ending cash balance (before borrowing) is $200,000, they will need to borrow $50,000 via a short-term line of credit. The interest rate on a short-term loan is 1% per month. At the end of each month, Widgetco earns interest of 0.5% on its beginning balance. Widgetco would like to have some idea about the maximum level of borrowing that will be needed during the January–June 1998 period. How will an increase or decrease (say of 20%) in the mean level of monthly sales influence the amount of financing needed? Assume that each month the previous month's line of short-term credit is "closed out" before a new loan is taken.

Solution Our work is in Figure 7.1 and the spreadsheet Loansim.wk4 or Loansim.xls. We will assume that the following sequence of events occurs during each month:

1 Widgetco observes its initial cash position.

2 It receives interest on initial cash position.

3 Receipts arrive, and Widgetco pays expenses (including closing out last month's short-term loan).

4 Widgetco takes out the current month's short-term loan.

5 It observes its final cash position (which becomes next month's initial cash position).

To simulate this situation, we proceed as follows:

Step 1 In the cell range C2:C4, we enter the percentage of sales from two months ago, one month ago, and the current month that will be received during the current month (Figure 7.1a).

Step 2 In cell C5 we enter the cost/sales price ratio (.80).

Step 3 In the cell range C12:K13 we enter the mean and standard deviation for monthly sales.

Step 4 In preparation for modeling a change in the mean level of monthly sales we enter the following statement in C6:

```
Excel:  =RISKSIMTABLE({.8,1,.1.2})
Lotus:  @RISKSIMTABLE(.8,1.1.2)
```

	A	B	C	D	E	F	G	H	I	J	K
1	**Cash budget**										
2	**Simulatior**	Percentage of 2 mo sale:	20%								
3		Percentage of 1 mo sale:	60%								
4		Percentage of current sa	20%								
5		Material costs/sales	80%								
6		Base Level	80%								
7		Interest rate on cash	0.5%								
8		Interest Rate on Loan	1.0%								
9											
10											
11			November	December	January	February	March	April	May	June	July
12		Mean sales	1500	1600	1800	1500	1900	2600	2400	1900	1300
13		Std Dev.	70	75	80	80	100	125	120	90	70
14		Actual Sales	1141.409	1133.23	1593.98	1164.664	1305.316	2193.457	1969.619	1517.212	1056.859
15		Receipts			1227.016	1415.967	1278.657	1454.814	1971.061	1923.905	
16		Initial Cash Balance			250	296.5346	419.7313	250	250	250	
17		Interest Earned			1.25	1.482673	2.098657	1.25	1.25	1.25	
18		Variable costs			931.7311	1044.252	1754.766	1575.695	1213.769	845.4875	
19		Fixed Costs			250	250	250	250	250	250	
20		Taxes					150			50	
21		Dividends								50	
22		Close out last month's loan				0	0	711.3214	1091.762	589.0524	
23		Cash Balance Before Loan			296.5346	419.7313	-454.279	-830.953	-333.22	390.6151	
24		Loan(if any)			0	0	704.2786	1080.953	583.2202	0	
25		Final Cash Position			296.5346	419.7313	250	250	250	390.6151	
26		Maximum Loan	1080.953								
27											
28					Variable Type						
29					Name	Maximum L	Maximum L	Maximum Loan/November			
30					Descriptior	Output(Si	Output(Si	Output(Sim#3)			
31					Cell	C26	C26	C26			
32					Minimum =	599.9026	558.5502	517.1979			
33					Maximum =	1421.177	1379.217	1337.257			
34					Mean =	993.745	952.5551	911.5671			
35					Std Deviat	152.2114	151.6789	151.2024			
36					Variance =	23168.32	23006.49	22862.16			
37					Skewness	-4.22E-02	-3.59E-02	-3.06E-02			
38					Kurtosis =	3.300999	3.298032	3.295034			
39					Errors Calc	0	0	0			
40					Percentile Values						
41					5% Perc =	733.8746	692.2736	650.9212			
42					10% Perc	795.8652	754.472	713.3547			
43					15% Perc	843.6269	805.5254	764.5151			
44					20% Perc	875.3325	833.9802	792.6278			
45					25% Perc	903.6876	863.4617	822.1094			
46					30% Perc	926.2383	884.886	843.5336			
47					35% Perc	947.5305	906.1782	864.8259			
48					40% Perc	956.0225	914.5223	873.17			
49					45% Perc	960.6771	919.3248	877.9724			
50					50% Perc	976.7765	935.4241	894.0718			
51					55% Perc	996.9088	959.917	920.4479			
52					60% Perc	1025.056	983.0959	941.6722			
53					65% Perc	1037.972	996.0118	954.416			
54					70% Perc	1066.358	1024.397	983.0223			
55					75% Perc	1087.939	1046.586	1005.234			
56					80% Perc	1115.839	1074.486	1033.134			
57					85% Perc	1153.675	1111.715	1069.993			
58					90% Perc	1206.223	1164.071	1122.18			
59					95% Perc	1236.026	1194.346	1152.994			
60					Filter Minimum =						
61		Filter Maximum =			Filter Maximum =						
62		Type(1 or 2) =			Type(1 or 2) =						
63		# Values Filtered =			# Values Filtered =						
64		Scenario #1 =			Scenario #	>75%	>75%	>75%			
65		Scenario #2 =			Scenario #	<25%	<25%	<25%			
66		Scenario #3 =			Scenario #	>90%	>90%	>90%			

Figure 7.1 Cash Budget Simulation

	A	B	C	D	E	F	G	H	I	J	K
1	**Cash budget**										
2	**Simulation**	Percentage of 2 mo sales	20%								
3		Percentage of 1 mo sales	60%								
4		Percentage of current sales	20%								
5		Material costs/sales	80%								
6		Base Level	80%								
7		Interest rate on cash	0.5%								
8		Interest Rate on Loan	1.0%								
9											
10											
11			November	December	January	February	March	April	May	June	July
12		Mean sales	1500	1600	1800	1500	1900	2600	2400	1900	1300
13		Std Dev.	70	75	80	80	100	125	120	90	70
14		Actual Sales	1141.409423	1133.2298	1593.979653	1164.663879	1305.315557	2193.457433	1969.618602	1517.211622	1056.859415
15		Receipts			1227.015695	1415.966527	1278.657369	1454.813596	1971.061292	1923.904972	
16		Initial Cash Balance			250	296.5345916	419.7313468	250	250	250	
17		Interest Earned			1.25	1.482672958	2.098656734	1.25	1.25	1.25	
18		Variable costs			931.7311034	1044.252445	1754.765947	1575.694881	1213.769298	845.4875319	
19		Fixed Costs			250	250	250	250	250	250	
20		Taxes					150			50	
21		Dividends								50	
22		Close out last month's loan				0	0	711.3213595	1091.762171	589.0523787	
23		Cash Balance Before Loan			296.5345916	419.7313468	-454.278574	-830.952644	-333.220177	390.6150615	
24		Loan(if any)			0	0	704.2785738	1080.952644	583.2201769	0	
25		Final Cash Position			296.5345916	419.7313468	250	250	250	390.6150615	
26		Maximum Loan	1080.952644								

Figure 7.1a Cash Budget Simulation

This statement will allow us to run three simulations. In the first simulation the mean of each month's sales will equal 80% of the mean given in Table 7.1. In the second simulation the mean of each month's sales will equal the mean given in Table 7.1. In the third simulation the mean of each month's sale will equal 120% of the mean given in Table 7.1.

Step 5 In C8 we enter the interest rate on short-term loans. In C7 we enter the interest rate earned on our cash balance.

Step 6 We assume that monthly sales are normally distributed. Then in cell C14 we generate our *actual* sales for November 1997 with the formula

```
Excel: =RISKNORMAL($C$6*C12,C13)

Lotus: @RISKNORMAL($C$6*C12,C13)
```

Copying this formula to the cell range D14:K14 generates actual sales for all other months of interest.

Step 7 In row 15 we operationalize (7.1) to generate monthly receipts. In cell E15 we compute January 1998 receipts with the formula

```
Excel: =$C$2*C14+$C$3*D14+$C$4*E14

Lotus: +$C$2*C14+$C$3*D14+$C$4*E14
```

Copying this formula to the cell range F15:J15 creates our receipts for the months February–June 1998.

Step 8 In cell E16 we enter our initial cash balance for January1998.

Step 9 In row 17 we compute the interest earned on each month's cash balance. In cell E17 we compute the interest earned on the January 1998 beginning cash balance with the formula

```
Excel: =$C$7*E16

Lotus: +$C$7*E16
```

Copying this formula to the cell range F17:J17 computes the interest payments for all other months.

Step 10 In row 18 we represent the fact that

(Current Month's Variable Costs) = .80*(Next Month's Sales)

To generate the variable costs for January 1998 we enter in cell E18 the formula

```
Excel: =$C$5*F14

Lotus: +$C$5*F14
```

Copying this formula to the cell range F18:J18 generates variable cost for the months February–June 1998.

Step 11 In rows 19–21 we enter the fixed costs, taxes, and dividends for each month.

Step 12 In row 22 we compute the amount of money (principal plus interest of 1%) needed to close out the previous month's short-term loan (if any). To compute the money needed in February to close out the January short-term loan we enter into cell F22 the formula

```
Excel:  =(1+$C$8)*E24

Lotus:  (1+$C$8)*E24
```

Note that the size of each month's loan will be computed in row 24.

Step 13 In row 23 we compute the cash balance before a short-term loan is taken out via the relationship

$$\begin{aligned}\text{(Cash Position Before Loan)} = &\ \text{(Beginning Cash Position)}\\ &+ \text{(Month's Receipts)}\\ &+ \text{(Interest on Beginning Cash Balance)}\\ &- \text{(Month's Expenses)}\\ &- \text{(Cost of Closing Out Last Month's Short-Term Loan)}\end{aligned} \tag{7.2}$$

To compute our January 1998 cash balance before a short-term loan is taken out we enter in cell E23 the formula

```
Excel:  =SUM(E15:E17)-SUM(E18:E22)

Lotus:  @SUM(E15:E17)-SUM(E18:E22)
```

Copying this formula to the cell range F23:J23 computes the cash balance for months February–June 1998 before the month's loan is taken out.

Step 14 In row 24 we compute the size of the loan (if any) taken out. To bring our ending cash balance to $250,000 we need to take out a loan if the value in row 23 is less than $250,000. Otherwise, no loan is needed. Thus in cell E24 we compute our January 1998 short-term loan with the formula

```
Excel:  =MAX(250-E23,0)

Lotus:  @MAX(250-E23,0)
```

Copying this formula to the cell range F24:J24 generates the short-term loan amounts for months February–June 1998.

In row 25 we compute each month's ending cash position by adding the amount of the loan to the cash position before the loan. Thus in cell E25 we compute our January ending cash position with the formula

```
Excel:  =E23+E24

Lotus:  +E23+E24
```

Copying this formula to the cell range F25:J25 computes the ending cash balance for months February–June.

Step 15 To compute the beginning cash balance for months February–June we enter in cell F16 the formula

```
Excel:  =E25

Lotus:  +E25
```

and copy this formula to the cell range F16:J16.

Step 16 Finally, we compute the maximum loan during the January–June period in cell C26 with the formula

```
Excel:  =MAX(E24:J24)

Lotus:  @MAX(E24:J24)
```

We now run three simulations with C26 as our output cell. The results (see Figure 7.1b) are pasted into Figure 7.1. We find that for the base sales scenario, there is a 5% chance that Widgetco will need to go at least $1,194,346 in debt. If average sales tend to be higher than anticipated, then there is a 5% chance that Widgetco will have to go $1,152,994 in debt. If average sales tend to be lower than anticipated, there is a 5% chance that the company will have to go at least $1,236,026 in debt.

This simulation also makes it easy to determine how changes in the collections of a company's accounts payable influence its financial requirements. For example, if 90% of all accounts receivable were collected during the month in which sales occurred, the company would find that its cash needs would be less.

Figure 7.1b
Cash Budget Simulation

Variable Type				
Name	Maximum Lo	Maximum Lo	Maximum Loan/November	
Description	Output (Sim#	Output (Sim#	Output (Sim#3)	
Cell	C26	C26	C26	
Minimum =	599.9026	558.5502	517.1979	
Maximum =	1421.177	1379.217	1337.257	
Mean =	993.745	952.5551	911.5671	
Std Deviation	152.2114	151.6789	151.2024	
Variance =	23168.32	23006.49	22862.16	
Skewness =	-4.22E-02	-3.59E-02	-3.06E-02	
Kurtosis =	3.300999	3.298032	3.295034	
Errors Calcul	0	0	0	
Percentile Values				
5% Perc =	733.8746	692.2736	650.9212	
10% Perc =	795.8652	754.472	713.3547	
15% Perc =	843.6269	805.5254	764.5151	
20% Perc =	875.3325	833.9802	792.6278	
25% Perc =	903.6876	863.4617	822.1094	
30% Perc =	926.2383	884.886	843.5336	
35% Perc =	947.5305	906.1782	864.8259	
40% Perc =	956.0225	914.5223	873.17	
45% Perc =	960.6771	919.3248	877.9724	
50% Perc =	976.7765	935.4241	894.0718	
55% Perc =	996.9088	959.917	920.4479	
60% Perc =	1025.056	983.0959	941.6722	
65% Perc =	1037.972	996.0118	954.416	
70% Perc =	1066.358	1024.397	983.0223	
75% Perc =	1087.939	1046.586	1005.234	
80% Perc =	1115.839	1074.486	1033.134	
85% Perc =	1153.675	1111.715	1069.993	
90% Perc =	1206.223	1164.071	1122.18	
95% Perc =	1236.026	1194.346	1152.994	
Filter Minimum =				
Filter Maximum =				
Type (1 or 2) =				
# Values Filtered =				
Scenario #1 =	>75%	>75%	>75%	
Scenario #2 =	<25%	<25%	<25%	
Scenario #3 =	>90%	>90%	>90%	

Problems

Group A

7.1 (Based on Altman 1986) You are to simulate the cash budget for a textile firm during the year 1999. We make the following assumptions (cash amounts are in thousands):

Monthly base sales are unknown, with a worst case estimate of 60,000, a best case estimate of 90,000, and a most likely value of 70,000.

Monthly Sales = (Monthly Base Sales) * (Seasonal Index for Month).

The seasonal indexes are given in Table 7.2.

Table 7.2

Jan	Feb	March	April	May	June	July	Aug	Sept	Oct	Nov	Dec
.85	.87	.90	.91	.94	1.02	1.03	1.06	1.06	1.10	1.12	1.13

Cash Sales During a Month = .1 * (Monthly Sales).

Monthly Collections for January and February = 50,882.

For other months Monthly Collections = .9 * (Sales from 2 Months Ago).

Monthly Total Cash Receipts = Cash Sales + Monthly Collections.

Together labor and material costs are a certain fraction of monthly sales (call this the cost factor). This fraction (the same for each month of the year) is normally distributed with $\mu = .70$ and $\sigma = .03$.

Monthly cash disbursements are given below:

Monthly Labor Costs = (Cost Factor) * .2 * (Monthly Sales).

Monthly Material Costs = 33,233 for January; otherwise Monthly Material Costs = (Cost Factor) * .8 * (Last Month's Sales).

Monthly Operating Expenses = .19 * (Monthly Base Sales).

Quarterly Interest Payments: in January the firm pays 2,780.

In April, July, and October

Quarterly Interest = Interest on Average Loan Balance for Last 3 Months (paid at short-term rate of interest for three months) + 3 Months of Interest on 96,000 Long-term Loan (paid at 8.5% annually).

Short-term interest rate for the year is normal with $\mu = 14\%$ and $\sigma = 1\%$.

Taxes are paid in January, April, July, and October. Each payment is 4,004.

Dividends of 2,000 are paid in February, May, August, and November.

Capital Expenditures are as follows:

March: 4,500 June: 4,600 Sept: 4,800 Dec: 4,900

At the beginning of January the cash balance = –7,015.

Desired Ending Cash for Month (after adjusting loan balance) = .188 * Monthly Sales.

Cash evolves as follows:

Beginning Cash for Month t = Desired Ending Cash for Month $t - 1$.

End of Month Cash for Month t = Beginning Cash for Month t + Net Cash Flow for Month t.

For all but January,

(Ending Loan Balance for Month t) = (Ending Loan Balance for Month $t - 1$) + (Desired Cash for Month t) – (End of Month Cash for Month t).

For January,

Ending Loan Balance = (Ending January Desired Cash Balance) – (January End of Month Cash).

Simulate the firm's yearly cash budget 500 times. Find the mean and variance of the firm's total net cash flow for 1999 as well as the mean and variance of the firm's monthly average loan balance for 1999. What is the probability that the firm's 1999 net cash flow will be negative? What is the probability that the firm's ending loan balance will be negative?

Note: Monthly Net Cash Flow = Monthly Cash Receipts – Monthly Cash Disbursements

7.2 (Based on Benninga 1989) You are the chief financial officer (CFO) for Carco, a small car rental company. You are trying to get some idea of what Carco's financial and income statements will look like during the current year (Year 0) and the next five years. The following relationships hold:

Current Assets for Each Year = CA * (Year's Sales)

where the CA for each year are independent normal random variables with mean .15 and standard deviation .02.

For each year

Fixed Assets at Cost = Depreciation + Net Fixed Assets.

Year 0 Depreciation = 330.

For $t \geq 1$, Year t Depreciation = Year $(t - 1)$ Depreciation + .1 * (Fixed Assets at Cost for Year $t - 1$).

Net Fixed Assets for Year t = (Year t sales) * (NFA), where the NFA for each year are independent normal random variables with mean .77 and standard deviation .04.

Total Assets for Each Year = Net Fixed Assets + Current Assets.

Current Liabilities for Each Year = (Year's Sales) * CL, where the CL are independent normal random variables with mean .08 and standard deviation .01.

Long-Term Debt for Year 0 = 280.

For each year, Carco wants to have the debt/equity ratios given in Table 7.3.

Table 7.3

Year 1	Year 2	Year 3	Year 4	Year 5
.48	.46	.44	.42	.40

Thus for $t \geq 1$,

Long-Term Debt for Year t = (Year t Debt/Equity Ratio) * (Year t Retained Earning + Year t Stock).

Year 0 Stock = 450.

Year t Stock = Year $t - 1$ Stock + Year t New Stock.

Year 0 Retained Earnings = 110.

For $t \geq 1$,

Year t Retained Earnings = Year $t - 1$ Retained Earnings + Year t Retention.

For each year,

Total Liabilities = Current Liabilities + Long-Term Debt + Stock + Retained Earnings.

The amount of new stock issued each year must be enough to make Total Assets = Total Liabilities.

The interest rate on current debt is 10.5%, while the interest rate on new debt is 9.5%. During each of the next five years, 20% of the current 280 in long-term debt is to be paid off. Then the total amount of new debt during year t is found from

Amount of Initial Debt Still Left + Total Year t New Debt = Year t Long-Term Debt.

New Debt for Year t = Total New Debt for Year t – Total New Debt for Year $t - 1$.

Year 0 Sales = 1000.

For $t \geq 1$,

Year t Sales = (Year $t - 1$ Sales) * SF,

where the SF are independent normal random variables with mean 1.1 and standard deviation .05. By the way, this does not mean that sales during successive years are independent.

Year t Expenses = (Year t Sales) * EF,

where the EF are independent normal random variables, with mean .80 and standard deviation .06.

To compute yearly interest payments remember that interest is 10.5% on old debt and 9.5% on new debt.

Depreciation for Year 0 = 110.

For $t \geq 1$,

Year t Depreciation = .10*(Year $t - 1$ At-Cost Fixed Asset).

For each year,

Before-Tax Profit = Sales – Expenses – Interest Payments – Depreciation.

Tax Rate = 47%.

Dividends for Each Year = 70% of After-Tax Profits.

Retention for Each Year = 30% of After-Tax Profits.

Set up a spreadsheet to model the current year (0) and next five years (1–5) of Carco's financial future. Now simulate the firm's future 500 times. Use your output to complete these statements and answer these questions:

a There is only a 5% chance that total new debt will exceed _____.

b On the average, total interest payments for the next five years will equal _____.

c What is the probability of profit being negative during Year 5?

Note: Your spreadsheet contains circular references. There are many of these; for example, stock purchased each year can be shown to depend on long-term debt, and long-term debt can be shown to depend on stock. To resolve the circular references proceed as follows:

Lotus: Use the commands /Worksheet Global Recalculate Iterations (then enter 20). This will ensure that the spreadsheet will recalculate itself 20 times. This recalculation will ensure that the values in your spreadsheet converge to their correct values.

Excel: Choose Options, Calculation Automatic, and then choose 20 Iterations.

7.3 You have been asked to simulate the cash inflows to a toy company for the next year. Monthly sales are independent random variables. Mean sales for the months January–March and October–December are $80,000, and mean sales for the months April–September are $120,000. The standard deviation for each month's sales is 20% of the month's mean sales. We model the method used to collect monthly sales as follows:

During each month a certain fraction of new sales will be collected. All new sales not collected become one month overdue.

During each month a certain fraction of one-month overdue sales is collected. The remainder become two months overdue.

During each month a certain fraction of two-month overdue sales is collected.

The remainder are written off as bad debts. You are given the following information from some past months.

Table 7.4

New Sales	New Sales Paid During Month
$40,000	$25,000
$50,000	$32,000
$60,000	$35,000
$20,000	$10,000
$30,000	$21,000
$45,000	$29,000

Table 7.5

One-Month-Old Sales	One-Month-Old Sales Paid Off During Month
$50,000	$26,000
$60,000	$33,000
$80,000	$36,000
$20,000	$8000
$25,000	$14,000
$30,000	$17,000
$40,000	$22,000

Table 7.6

Two-Month-Old Sales	Two-Month-Old Sales Collected
$30,000	$23,000
$40,000	$30,000
$50,000	$43,000
$60,000	$50,000
$70,000	$58,000
$40,000	$34,000

Using this information you should be able to build a simulation model that generates the total cash inflow for each month. You do not need to run any regressions. Just develop a simple forecasting model and build into the simulation somewhere the error of your forecasting model. Assuming that during January there are $120,000 of one-month-old sales outstanding and $140,000 of two-month-old sales outstanding, you are 95% sure that total cash inflow for the year will be between ____ and _____.

7.4 Estimates of mean monthly sales for the months October 1995–March 1997 are as follows:

Table 7.7

	1995	1996	1997
January		$70,000	$130,000
February		$80,000	$130,000
March		$80,000	$140,000
April		$90,000	
May		$100,000	
June		$100,000	
July		$90,000	
August		$100,000	
September		$110,000	
October	$40,000	$120,000	
November	$50,000	$120,000	
December	$60,000	$130,000	

Your goal is to model Toyco's 1995 cash budget given the following assumptions:

At the beginning of each month Toyco wants to have a cash balance of at least $20,000, and it will borrow sufficient funds to achieve this goal.

All sales are for credit; 70% of all payments are collected in the first month after sale, 20% in the second month after sale, and 10% are collected in the third month after sale.

Inventory at the beginning of each month should equal forecasted sales for the next three months.

Merchandise purchased for sale incurs a cost equal to 75% of sales. On purchases made each month, 70% is paid in the first month after purchase and 30% is paid in the second month after purchase.

Selling and administrative expenses incurred during a month equal $8500 + .09*(Sales). These expenses are paid at a rate of 70% during the current month and 30% in the month following.

If money must be borrowed, it is borrowed month-to-month at an interest rate of 0.7%. Borrowing takes place in multiples of $1000. To model this situation use the ceiling function. For example, `CEILING(2500,1000) = 3000`. Note you will need an `IF` statement because `CEILING(-500,1000)` is not defined.

In February 1995 capital expenditures of $20,000 are incurred. In July and October capital expenditures of $30,000 are incurred.

Existing fixed assets are depreciated at $1000 per month. Additional capital expenses are depreciated on a straight-line basis at a rate of 1% per month, beginning in the month after the capital expense is incurred.

The sales for each month have a standard deviation of 15,000 about the forecasted sales.

a Use your model to predict average profit for 1995. How high is profit likely to go? How low?

b To be sure of having enough money to borrow, how big a line of credit would you need for 1995?

c Suppose you could ensure that all accounts receivable could be collected in the month after sale and that beginning inventory for each month could be trimmed to 80% of the next three month's forecasted sales. How would this new scenario change your answers to (a) and (b)?

References

Altman, E. 1986. *Handbook of Corporate Finance*. New York: Wiley.
Benninga, S. 1989. *Numerical Methods in Finance*. Cambridge: MIT Press.

CHAPTER 8

A Simulation Approach to Capacity Planning

We now show how to use simulation to analyze the capacity needed for a product (Figure 8.1). We use past data to forecast future sales and then come up with a capacity level that maximizes expected discounted profit. Finally, we introduce the concept of downside risk and determine a capacity level that minimizes downside risk.

Example 8.1 Wozac Capacity Example

Eli Daisy has taken over the production of Wozac from a rival drug company. Wozac's annual sales from 1985 to 1994 are given in Table 8.1.

Table 8.1

Year	Sales (thousands of units)
1985	500
1986	544
1987	593
1988	672
1989	723
1990	757
1991	848
1992	948
1993	964
1994	1011

	A	B	C	D	E	F	G	H	I	J	K	L
1	Year	Sales(thousands of units)			Figure 8.1							
2	1985	500				Interest rate	0.1					
3	1986	544	500			Unit variable cost	5.9					
4	1987	593	544			Sales price	10					
5	1988	672	593			Operating cost/unit plant capacity	1					
6	1989	723	672			Fixed Cost slope per unit of cap	10					
7	1990	757	723			Fixed Cost intercept	5000000					
8	1991	848	757									
9	1992	948	848			Variable cost slope per 100000 ca	0.1					
10	1993	964	948			Variable Cost intercept	6				Downside risk	$2,592,480.72
11	1994	1011	964			Capacity	1100000				NPV	$6,407,519.28
12						Demand	Sales	Sales Rev	Variable Cost	Fixed Cost	Operating Cost	Profit
13				1995		1116661.791	1100000	11000000	6490000	16000000	1100000	-12590000
14				1996		1230506.521	1100000	11000000	6490000		1100000	3410000
15				1997		1326307.074	1100000	11000000	6490000		1100000	3410000
16				1998		1370433.557	1100000	11000000	6490000		1100000	3410000
17				1999		1459570.868	1100000	11000000	6490000		1100000	3410000
18				2000		1501294.355	1100000	11000000	6490000		1100000	3410000
19				2001		1542929.936	1100000	11000000	6490000		1100000	3410000
20				2002		1555618.624	1100000	11000000	6490000		1100000	3410000
21				2003		1542425.989	1100000	11000000	6490000		1100000	3410000
22				2004		1626497.718	1100000	11000000	6490000		1100000	3410000
23												
24												
25						Simulation Results for T941.XLS						
26												
27						Iterations= 100						
28						Simulations= 7						
29						# Input Variables= 11						
30						# Output Variables= 2						
31						Sampling Type= Latin Hypercube						
32						Runtime= 00:01:34						
33												
34						Simulation Results for T941.XLS						
35												
36						Iterations= 100						
37						Simulations= 7						
38						# Input Variables= 11						
39						# Output Variables= 2						
40						Sampling Type= Latin Hypercube						
41						Runtime= 00:02:15						
42												
43						Summary Statistics						
44												
45						Cell	Name	Minimum	Mean	Maximum		
46												
47						L10	(Sim#1) Downside ris	2592481	2783408	4099239		
48						L10	(Sim#2) Downside ris	1126825	1880743	4083012		
49						L10	(Sim#3) Downside ris	83840.3	1483336	4309801		
50						L10	(Sim#4) Downside ris	0	1560449	4847479		
51						L10	(Sim#5) Downside ris	0	2000263	5661346		
52						L10	(Sim#6) Downside ris	0	2646793	6501842		
53						L10	(Sim#7) Downside ris	0	3371301	7342338		
54						L11	(Sim#1) Downside ris	4900761	6216592	6407520		
55						L11	(Sim#2) Downside ris	4916989	7119257	7873175		
56						L11	(Sim#3) Downside ris	4690200	7516664	8916160		
57						L11	(Sim#4) Downside ris	4152521	7457208	9501291		
58						L11	(Sim#5) Downside ris	3338655	7029042	9810766		
59						L11	(Sim#6) Downside ris	2498159	6370922	9623687		
60						L11	(Sim#7) Downside ris	1657663	5631198	9125507		

Figure 8.1 Simulation Output for Capacity Planning

Daisy must build a plant to produce Wozac by the beginning of 1995. Once the plant is built, the plant's capacity cannot be changed. Each unit sold brings in \$10 in revenue.

The fixed cost (in dollars) of producing a plant that can produce x units per year of the drug is

Fixed Cost of Building Plant = 5,000,000 + 10x.

This cost is assumed to be incurred at the end of 1995. We assume that all cost and sales cash flows are incurred at the end of each year.

If a plant of capacity x is built, the variable cost of producing a unit of Wozac will be

Variable Cost per unit = 6 – .1 * (x – 1,000,000)/100,000.

Thus a plant capacity of 1,100,000 units will result in a variable cost of \$5.90.

Each year a plant operating cost of \$1 per unit of capacity is also incurred.

If demand for a year exceeds production capacity, all sales in excess of plant capacity are assumed lost. Determine a capacity level that will maximize expected discounted (at an interest rate of 10%) profits for the time period 1995–2004.

Solution We begin by developing a model to forecast future annual demand for Wozac. We postulate that for some constants B_0 and B_1,

$$\text{Demand for Wozac During Year } t = (B_0) + (B_1) * (\text{Demand for Wozac during year } t - 1) + \text{Error Term} \quad \textbf{(8.1)}$$

To estimate (8.1) we ran a regression with last year's sales as the independent variable and this year's sales as the dependent variable. To do this we used the Excel Regression capability (under Data Analysis Tools) or the Lotus Regression capability (under Range Analyze Regression). Our x-range was C3:C11 and our y-range was B3:B11. The results of the regression are in Figure 8.2.

Next we estimate B_0 to be the intercept of this regression (in cell O18), B_1 to be the slope of this regression (in cell O19), and the error term to be a normal random variable with mean 0 and standard deviation equal to the standard error of the regression (in cell O8). Thus our model for future annual demand (in thousands of units) is

$$\text{Year } t \text{ Demand} = 67.43 + .985 * (\text{Year } t - 1 \text{ Demand}) + \text{Normal RV with Mean 0 and Std. Dev. } 29.32 \quad \textbf{(8.2)}$$

Note that this model does not assume that demand during successive years is independent. This is because Year t – 1 demand strongly influences Year t demand.

Entering Simulation Inputs

We now enter relevant information in the cell range G2:G11 (see Figure 8.1a).

	N	O	P	Q	R	S	T	U	V
1	**Regresssion Output Figure 8.2**								
2	SUMMARY OUTPUT								
3									
4	*Regression Statistics*								
5	Multiple R	0.98659348							
6	R Square	0.97336669							
7	Adjusted R S	0.96956194							
8	Standard Err	29.3232556							
9	Observations	9							
10									
11	ANOVA								
12		*df*	*SS*	*MS*	*F*	*Significance F*			
13	Regression	1	219975.249	219975.249	255.828808	9.06711E-07			
14	Residual	7	6018.973216	859.853317					
15	Total	8	225994.2222						
16									
17		*Coefficients*	*Standard Error*	*t Stat*	*P-value*	*Lower 95%*	*Upper 95%*	*Lower 95.0%*	*Upper 95.0%*
18	Intercept	67.4333839	45.88142549	1.46973167	0.18509355	-41.0588698	175.9256377	-41.0588698	175.9256377
19	X Variable 1	0.98535647	0.061605381	15.9946494	9.0671E-07	0.839682999	1.131029948	0.839682999	1.131029948

Figure 8.2 Regression Output for Capacity Planning

Step 1 In cell G2 we enter the interest rate (10%).

Step 2 In cell G11 we enter possible capacity levels with the formula

```
=RISKSIMTABLE({1100000,1200000,1300000,1400000,
1500000,1600000,1700000})
```

This allows @RISK to collect statistics on capacity levels ranging from 1.1 million to 1.7 million.

Step 3 After entering the variable unit cost intercept of 6 in cell G10 and the variable unit cost slope of .1 in cell G9, we compute in cell G3 the unit variable cost of production with the formula

```
=G10-G9*(G11-1000000)/100000
```

Step 4 In cell G4 we enter the sales price for a unit of Wozac ($10). We have not allowed for any price increases.

Step 5 In cell G5 we enter the $1 annual fixed cost/unit of capacity.

Step 6 In cell G6 we enter the slope (10) of the cost of building plant capacity.

Step 7 In cell G7 we enter the intercept (5,000,000) for the cost of building plant capacity.

Step 8 In rows 13–22 we generate cash flows for each of the years 1995–2004. We generate (random) demands for each year in the cell range F13:F22.

	A	B	C	D	E	F	G	H	I	J	K	L
1	Year	Sales(thousands of units)			Figure 8.1							
2	1985	500				Interest rate	0.1					
3	1986	544	500			Unit variable cost	5.9					
4	1987	593	544			Sales price	10					
5	1988	672	593			Operating cost/unit plant capacity	1					
6	1989	723	672			Fixed Cost slope per unit of cap	10					
7	1990	757	723			Fixed Cost intercept	5000000					
8	1991	848	757									
9	1992	948	848			Variable cost slope per 100000 cap	0.1					
10	1993	964	948			Variable Cost intercept	6				Downside risk	$2,592,480.72
11	1994	1011	964			Capacity	1100000				NPV	$6,407,519.28
12						Demand	Sales	Sales Rev	Variable Cost	Fixed Cost	Operating Cost	Profit
13				1995		1116661.791	1100000	11000000	6490000	16000000	1100000	-12590000
14				1996		1230506.521	1100000	11000000	6490000		1100000	3410000
15				1997		1326307.074	1100000	11000000	6490000		1100000	3410000
16				1998		1370433.557	1100000	11000000	6490000		1100000	3410000
17				1999		1459570.868	1100000	11000000	6490000		1100000	3410000
18				2000		1501294.355	1100000	11000000	6490000		1100000	3410000
19				2001		1542929.936	1100000	11000000	6490000		1100000	3410000
20				2002		1555618.624	1100000	11000000	6490000		1100000	3410000
21				2003		1542425.989	1100000	11000000	6490000		1100000	3410000
22				2004		1626497.718	1100000	11000000	6490000		1100000	3410000

Figure 8.1a

To generate our 1995 demand we operationalize our regression model (8.1) by entering in cell F13 the formula
`=1000*RISKNORMAL($O$18+$O$19*B11,$O$8)`

To generate 1996 demand we enter in cell F14 the formula
`=1000*RISKNORMAL($O$18+($O$19*F13/1000),$O$8)`

Copying this formula to the cell range F15:F22 generates demands for the years 1997–2004.

Step 9 In the cell range G13:G22 we generate annual sales via the relationship

Annual Sales = Min(Annual Demand, Capacity Level)

Thus, year 1995 sales are computed in cell G13 with the formula `=MIN(F13,$G$11)`.

Copying this formula to the cell range G14:G22 computes annual sales for the years 1996–2004.

Step 10 In the cell range H13:H22 we compute annual sales revenue by simply multiplying annual sales units by the sales price of $10. We compute 1995 sales revenue in cell H13 with the formula `=$G$4*G13`.

Copying this formula to the cell range H14:H22 computes sales revenues for the years 1996–2004.

Step 11 In the cell range I13:I22 we compute annual variable production cost for each year by multiplying annual sales units times the variable cost of production. We compute 1995 variable costs in cell I13 with the formula `=$G$3*G13`.

Copying this formula to the cell range I14:I22 computes our annual variable costs for the years 1996–2004.

Step 12 In cell J14 we enter the fixed cost of building the plant with the formula `=$G$7+$G$6*$G$11`.

Step 13 In the cell range K13:K22 we compute the annual plant operating costs (exclusive of variable and one-time building costs).

We compute the 1995 annual plant operating costs in cell K13 with the formula `=$G$5*$G$11`.

Copying this formula to the cell range K14:K22 computes the annual plant operating costs for the years 1996–2004.

Step 14 In the cell range L13:L22 we compute each year's profit. In cell L13 we compute the 1995 profit with the formula `=H13-I13-J13-K13`.

Copying this formula to the cell range L14:L22 computes the profit earned during the years 1996–2004.

Step 15 We now compute the NPV of all our cash flows in cell L11 with the formula `=NPV(G2,L13:L22)`.

Note that this formula assumes cash flows for each year occur at the end of the year.

Running the Simulation

We now run seven simulations (one for each capacity level) with cell L11 as our output cell (See Figure 8.1b). We find that the capacity level that maximizes expected discounted NPV is 1.3 million.

Suppose that Daisy has set a target level of $9 million for the discounted profit earned by Wozac during the years 1995–2004. One way to model this objective is to assume that if we fall short of our target by $\$x$, then a penalty of $\$x$ is incurred. We call this penalty the **downside risk** associated with the target profit level. In cell L10 we compute the downside risk with the formula `=max(9000000-L11,0)`.

Note that this formula yields no penalty (and no credit!) when our target is met. Running seven simulations again shows that an annual capacity of 1.3 million minimizes expected downside risk. It is purely coincidental that in this case the expected profit and the downside risk criteria yield the same optimal capacity level.

Simulation Results for T941.XLS				
Iterations= 100				
Simulations= 7				
# Input Variables= 11				
# Output Variables= 2				
Sampling Type= Latin Hypercube				
Runtime= 00:02:15				
Summary Statistics				
Cell	Name	Minimum	Mean	Maximum
L10	(Sim#1) Downside risl	2592481	2783408	4099239
L10	(Sim#2) Downside risl	1126825	1880743	4083012
L10	(Sim#3) Downside risl	83840.3	1483336	4309801
L10	(Sim#4) Downside risl	0	1560449	4847479
L10	(Sim#5) Downside risl	0	2000263	5661346
L10	(Sim#6) Downside risl	0	2646793	6501842
L10	(Sim#7) Downside risl	0	3371301	7342338
L11	(Sim#1) Downside risl	4900761	6216592	6407520
L11	(Sim#2) Downside risl	4916989	7119257	7873175
L11	(Sim#3) Downside risl	4690200	7516664	8916160
L11	(Sim#4) Downside risl	4152521	7457208	9501291
L11	(Sim#5) Downside risl	3338655	7029042	9810766
L11	(Sim#6) Downside risl	2498159	6370922	9623687
L11	(Sim#7) Downside risl	1657663	5631198	9125507

Figure 8.1b

Problems

Group A

8.1 Our regression model (8.2) implies that average demand for our product is increasing by around 67,000 units per year. A look at the data indicates that the rate of increase in sales is increasing as sales increase. This could be modeled by assuming sales grow annually by an average percentage. Thus a better model for sales might be

$$\text{Year } t \text{ Sales} = (\text{constant}) * (\text{Year } t - 1 \text{ Sales}) * (\text{Error Term}) \quad \textbf{(8.3)}$$

To estimate the constant, simply take the average percentage annual increase in sales as an estimate of the constant. Then for each year t I would estimate

$$\text{Error } t = (\text{Actual Year } t \text{ Sales}) / (\text{Predicted Year } t \text{ Sales})$$

To simulate future demand in @RISK I would now assume that the error term in (8.3) is normally distributed with a mean equal to the average of the Error t and a standard deviation equal to the standard deviation of the Error t.

a Determine the optimal capacity level using the data above.

b Which model appears more appropriate, (8.3) or (8.2)?

Group B

8.2 Barbella dolls sell for $15 per doll. Annual sales (in tens of thousands) and production costs for the last eleven years are given in Table 8.2. We assume that costs increase an average of 3% per year and that each year units produced equal units sold.

Table 8.2

Year	Sales	Production Costs (millions of dollars)
1	500	12.00
2	570	11.90
3	610	11.03
4	650	13.36
5	720	14.14
6	750	13.76
7	790	14.65
8	870	18.13
9	890	18.24
10	960	18.24
11	990	20.48

Complete the following statements.

a There is a 5% chance that next year's profits will be below _____.

b There is a 5% chance that next year's profits will be above _____.

c The most likely value for next year's profit is _____.

8.3 Bat Corporation produces one-pound salmon cans for the Canadian salmon industry. Each year the salmon spawn during a 24-hour period and must be immediately canned. Bat Corporation has the following agreement with the salmon industry. Bat delivers the number of cans Bat desires. Then the salmon are caught. For each can by which Bat falls short of the salmon industry's needs, Bat pays the salmon industry a $2.00 penalty. Cans cost Bat $1.00 to produce and are purchased for $2.00 per can. If any cans are leftover, they are returned to Bat (Bat must then refund the $2.00), and the cans are put in storage for next year. Each year a can is held in storage, a carrying cost equal to 20% of the can's production cost is incurred. It is well known that the number of salmon harvested during a year is strongly related to the number of salmon harvested the previous year. This is borne out by the data in Table 8.3.

Table 8.3

Year	Salmon Cans
1	600
2	464
3	364
4	334
5	383
6	399
7	417
8	500
9	537
10	622

What production strategy should Bat follow? Hint: Consider strategies of this form—for some x, produce enough cans for Year t to bring on-hand inventory to x + (Predicted Year t Can Requirements) and simulate, say a 20-year planning horizon. For example, suppose $x = 100{,}000$. If we predict 500,000 cans will be needed and we have 80,000 cans leftover, we need to produce 500,000 + 100,000 – 80,000 = 520,000 cans.

CHAPTER 9

Simulation and Bidding

In situations in which you must bid against competitors, simulation can often be used to determine the correct bid. Usually you do not know what a competitor will bid, but you may have an idea about the range of bids a competitor may choose. In this chapter we show how to use simulation to determine a bid that maximizes your expected profit. Before discussing our example, we must briefly discuss generating observations from a uniformly distributed random variable.

9.1 Uniform Random Variables

A random variable is said to be **uniformly distributed** on the closed interval $[a, b]$ (written $U(a, b)$) if the random variable is equally likely to assume any value between a and b inclusive. To generate samples from a $U(a, b)$ random variable enter the formula

```
Excel: =RISKUNIFORM(a,b)

Lotus: @RISKUNIFORM(a,b)
```

into a cell.

9.2 A Bidding Example

We now show how to use simulation to determine a bid that maximizes expected profit.

Example 9.1

You are going to make a bid on a construction project. You believe it will cost you $10,000 to complete the project. Four potential competitors are going to bid against you. Based on past history, you believe that each competitor's bid is equally likely to be any value between your cost of completing the project and triple your cost of completing the project. You also believe that each competitor's bid is independent of the other competitors' bids. What bid maximizes your expected profit?

Solution In our solution all amounts will be in thousands of dollars. The statement of the problem implies that each competitor's bid is $U(10, 30)$, and the bids of the competitors are independent. Our simulation is shown in Figure 9.1 (file Bid.wk4 or Bid.xls). We proceed as follows:

Step 1 In cell C3 we enter the cost of the project.

Step 2 In cell C4 we enter ten possible bids (11, 12, 13, 14, 15, 16, 17, 18, 19, and 20) with the formula

```
Excel: =RISKSIMTABLE({11,12,13,14,15,16,17,18,
19,20})

Lotus: @RISKSIMTABLE(11,12,13,14,15,16,17,18,
19,20)
```

Step 3 In C5 I generate the bid of my first competitor by entering the formula

```
Excel: =RISKUNIFORM(C$3,3*C$3)

Lotus: @RISKUNIFORM(C$3,3*C$3)
```

Copying this formula to the range C6:C8 generates the bids of the other three competitors. (Why does this ensure that their bids are independent?)

Step 4 In cell C9 we compute the actual profit for this trial by entering the formula

```
Excel: =IF(C4<=MIN(C5:C8),C4-C3,0)

Lotus: @IF(C4<=@MIN(C5..C8),C4-C3,0)
```

This ensures that if we win the bid (`C4<=@MIN(C5..C8)`), then our profit equals our bid less the project cost of $10,000; if we don't win the

	A	B	C	D	E	F
1	**Bidding Example**		**Figure 9.1**			
2						
3	My cost(thousands)		10			
4	My bid(thousands)		11			
5	Competitor 1 Bid		14.8619422			
6	Competitor 2 Bid		24.9901557			
7	Competitor 3 Bid		14.31341778			
8	Competitor 4 Bid		17.86299302			
9	Profit(thousands)		1			
10						
11						
12						
13		Iterations= 400				
14		Simulations= 10				
15		# Input Variables= 5				
16		# Output Variables= 1				
17		Sampling Type= Latin Hypercube				
18		Runtime= 00:02:45				
19						
20		Summary Statistics				
21						
22		Cell	Name	Minimum	Mean	Maximum
23						
24		C9	(Sim#1) Profit(thousa...	0	0.81	1
25		C9	(Sim#2) Profit(thousa...	0	1.3	2
26		C9	(Sim#3) Profit(thousa...	0	1.575	3
27		C9	(Sim#4) Profit(thousa...	0	1.66	4
28		C9	(Sim#5) Profit(thousa...	0	1.5625	5
29		C9	(Sim#6) Profit(thousa...	0	1.41	6
30		C9	(Sim#7) Profit(thousa...	0	1.155	7
31		C9	(Sim#8) Profit(thousa...	0	0.96	8
32		C9	(Sim#9) Profit(thousa...	0	0.7425	9
33		C9	(Sim#10) Profit(thous...	0	0.55	10

Figure 9.1
Bidding Simulation

bid (`C4>@MIN(C5..C8)`), then we earn no profit. This statement assumes that we win all ties, but the chance of a tie bid is negligible (why?), so this really does not matter. To see how things work, hit the recalculation (F9) button and see how the cells of the spreadsheet change.

Step 5 To determine the bid that maximizes expected profit we ran 400 iterations of this spreadsheet for each bid with @RISK. From Figure 9.1 it appears that a bid between \$13,000 and \$15,000 will maximize expected profit (with an expected profit of \$1660).

Step 6 To zero in on the bid that maximizes expected profit we replaced the formula in cell C4 with

```
Excel: =RISKSIMTABLE({13.2,13.4,13.6,13.8,14,
14.2,14.4,14.6,14.8})

Lotus: @RISKSIMTABLE(13.2,13.4,13.6,13.8,14,
14.2,14.4,14.6,14.8)
```

100 iterations of this spreadsheet indicate that a bid of around $14,200 maximizes expected profit (an expected profit of around $1800 is earned).

Problems

Group A

9.1 If the number of competitors were to double, how would the optimal bid change?

9.2 If the average bid for each competitor stayed the same, but their bids exhibited less variability, would the optimal bid increase or decrease? To study this question assume that each competitor's bid follows each of the following random variables:

a $U(15,25)$

b $U(18,22)$

9.3 Warren Millken is attempting to take over Biotech Corporation. The worth of Biotech depends on the success or failure of several drugs under development. Warren does not know the *actual* (per share) worth of Biotech, but the current owners of Biotech do know the actual worth of the company. Warren assumes that Biotech's actual worth is equally likely to be between $0 and $100 per share. Biotech will accept Warren's offer if it exceeds the true worth of the company. For example, if the current owners think Biotech is worth $40 per share and Warren bids $50 per share, they will accept the bid. If the current owners accept Warren's bid, then Warren's corporate strengths immediately increase Biotech's market value by 50%. How much should Warren bid?

Group B

9.4 You are bidding for a project that will cost you $20,000. In Table 9.1 you are given the past bids of your two competitors. You can use this data and regression to come up with a way to simulate the competition's bids. If you want to maximize your expected profit from this contract, what should your bid be (within $1000)? Run 400 trials for each bid you are considering.

Hint: Assume that for each competitor, their bid is given by $B_0 + B_1$*(Your Cost) + (Error Term). Assume that the error term is normally distributed with mean 0. You can use regression to estimate B_0, B_1, and the standard deviation of the error term.

Table 9.1

Our Cost of Project	Actual Bid by Competitor 1	Actual Bid by Competitor 2
$10,000	$10,920	$14,920
$14,000	$16,470	$19,640
$16,000	$25,230	$19,450
$18,000	$23,120	$18,110
$30,000	$39,990	$46,560
$25,000	$37,380	$38,350
$38,000	$56,150	$53,530
$44,000	$60,210	$48,050
$24,000	$38,070	$30,920

9.5 In Example 9.1 suppose competitors do not bid on all projects. More specifically, there is only a 40% chance that a given competitor will bid against you on a given project. Determine the bid that maximizes expected profit.

Hint: Use the @RISK `RISKBINOMIAL` function to generate the number of competitors that are actually bidding. To model the competitors who do not place a bid, use an `@IF` statement to ensure that their "bid" is $30,000. (Why $30,000?)

CHAPTER 10

Deming's Funnel Experiment

Edwards Deming (1900–1993) was an American statistician whose views on quality management revolutionized the way companies do business across the world. Deming has been given much of the credit for Japan's spectacular post–World War II economic recovery.

Until his death, Deming toured the U.S., giving a famous four-day seminar on quality management. After attending his seminar, many U.S. companies (including Xerox, GM, and Ford) reorganized their business to reflect Deming's management philosophy as embodied in his famous 14 points. In fact, GM's Saturn plant is run almost completely in accordance with Deming's 14 points. We strongly recommend that the reader pick up a copy of Deming's *Out of the Crisis*. It's terrific!

An important component of Deming's seminar was his famous funnel experiment. The funnel experiment was designed to show how businesses often overadjust stable processes. To illustrate the idea, suppose we are in the business of drilling a hole in the center of a 2″-square piece of wood. Suppose that in the past the holes we have drilled were, on average, in the center of the wood and the x- and y-coordinates had a standard deviation of .1″. Also suppose that the drilling process has been stable. Essentially, this means that *at any point in time*, our holes are, on average, in the center of the square and the deviation (measured for both the x- and y-coordinates) from the center of the square follows a normal distribution with mean 0 and standard deviation .1″. This means that for 68% of the holes, the x-coordinate is within .1″ of the center; for 95% of the holes, the x-coordinate is within .2″ of the center; and for 99.7% of the holes, the x-coordinate is within .3″ of the center. This gives a picture of the variability inherent in our drilling process. Suppose a hole is drilled and its x- and y-coordinates are x = 1.1″ and y = 1″ (the center of the square is x = 1″ and y = 1″). A natural reaction is to reduce (if possible) the x-setting of the drill by .1″ to correct for the fact that the x-coordinate was too high. Deming's funnel experiment shows that this approach to adjusting a process will greatly increase the variability of both the x- and y-coordinates of the position where the hole is drilled!

To illustrate this idea, Deming placed a funnel above a small target on a floor and dropped marbles through the funnel in an attempt to hit the target. Of course, many balls did not hit the target. Our goal is to minimize the average distance by which the dropped balls miss the target. This is equivalent to minimizing the variability of the x- and y-coordinates of the position where the ball lands. Deming proposed four rules that can be used to adjust the funnel.

Rule 1: Never move the funnel.

Rule 2: After each ball is dropped, move the funnel (*relative to its last position*) to compensate for any error. To illustrate, suppose the target has coordinates (0, 0) and the funnel begins directly over the target. If on the first drop the ball lands at (.5, .1) we compensate by repositioning the funnel at (0 – .5, 0 – .1) = (–.5, –.1). If the second drop has coordinates (1, 2), we now reposition the funnel at (– .5 – 1, –.1 – 2) = (– 1.5, – 2.1).

Rule 3: Move the funnel [*relative to its original position of (0,0)*] to compensate for error. If on the first drop the ball lands at (.5, .1) we compensate by repositioning the funnel at (0 – .5, 0 – .1) = (– .5, – .1). If the second drop has coordinates (1, 2), we now reposition the funnel at (0 – 1, 0 – 2) = (– 1, – 2).

Rule 4: To minimize variability, always reposition the funnel *directly over the last drop*. Thus if the first ball lands at (.5, 1) we reposition the funnel to (.5, 1). If the second drop has coordinates (1, 2), we reposition the funnel to (1, 2).

To evaluate these rules we make the following assumption: *The x-coordinate on each drop is normally distributed with a mean equal to the x-coordinate of the funnel and a unit variance.* A similar statement holds for the y-coordinate. This models the uncertainty or variability inherent in dropping the balls.

We now set up a spreadsheet to simulate each rule. In each spreadsheet we used @RISK to simulate 50 drops of the ball. For Rule 1, we kept track of the position of the 50th ball for 400 iterations. Then we recorded the mean and variance of the x- and y-coordinates of the 50th ball for these 400 drops. For Rule 2, we ran 900 iterations and kept track of the position of the 50th ball. A good rule should have an average x- and y-coordinate of 0 and a small variance for each coordinate.

10.1 Simulating Rule 1 (Don't Touch That Funnel!)

Figure 10.1 (file Dem1.wk4 or Dem1.xls) simulates 50 drops with Rule 1. We proceed as follows:

Step 1 In C2 and D2 we enter 0s to represent the initial location of the funnel.

Step 2 Copy these 0s to C3:D51 to ensure that the funnel is never moved.

Step 3 In A2 we enter the formula

```
Excel: =RISKNORMAL(C2,1)

Lotus: @RISKNORMAL(C2,1)
```

to generate the x-coordinate for the first drop. Copying this formula to A2:B51 will generate the x- and y-coordinates for all 50 drops.

Using @RISK we simulated the x-coordinate and y-coordinate of the 50th drop 400 times. We found the average x-coordinate to be .0006 with a variance of .995. The average y-coordinate of the 50th drop is .0003 with a variance of .991. Clearly with no movement of the funnel we would expect an average x-coordinate of 0 and a variance of 1 on the 50th ball. Our simulation yielded results very close to this.

10.2 Simulating Rule 2

In Figure 10.2 (file Dem2.wk4 or Dem2.xls) we simulated Rule 2. We proceed as follows:

Step 1 To begin we enter 0s in cells C2 and D2 to represent the initial position of the funnel. In A2 we generated the x-coordinate for the first drop by entering the formula

```
Excel: =RISKNORMAL(C2,1)

Lotus: @RISKNORMAL(C2,1)
```

Copying this formula to B2 generates the y-coordinate for the first drop.

Step 2 In C3 we compute the x-coordinate of the funnel for the second drop by correcting for the error (relative to current position) on the first drop. The first drop had an x-coordinate of A2, so we need to adjust the current x-coordinate of the funnel by –A2 to compensate for the error. This gives us a new x-coordinate for the funnel of

```
Excel: =C2-A2

Lotus: +C2-A2
```

Figure 10.1
Deming Funnel Experiment: Rule 1

	A	B	C	D	E	F	G	H
1	DROPX	DROPY	POSX	POSY	**Deming Rule 1**		Figure 10.1	
2	-0.167	1.31037	0	0	1			
3	-0.6232	-1.3513	0	0	2		/DROPX	/DROPY
4	-1.6718	-0.285	0	0	3		Output	Output
5	-0.1053	-1.9362	0	0	4		A51	B51
6	-0.1967	0.48809	0	0	5	MINIMUM	-2.96221	-2.8138
7	-0.0768	-0.0132	0	0	6	MAXIMUM	2.924879	2.894287
8	-0.5971	-1.9951	0	0	7	MEAN	5.83E-04	2.57E-04
9	-0.2662	2.24548	0	0	8	STD DEV	0.997721	0.995738
10	-0.3191	0.7787	0	0	9	VARIANCE	0.995447	0.991494
11	1.352	0.17846	0	0	10			
12	1.28224	1.18444	0	0	11			
13	-0.6678	0.56922	0	0	12			
14	0.92466	0.30994	0	0	13			
15	0.73797	0.35842	0	0	14			
16	-1.6211	1.35794	0	0	15			
17	0.50104	-0.0595	0	0	16			
18	1.37073	0.43578	0	0	17			
19	0.98776	2.52877	0	0	18			
20	-0.7245	1.17321	0	0	19			
21	0.19211	0.6847	0	0	20			
22	-0.2151	0.08514	0	0	21			
23	-3.1387	-1.0106	0	0	22			
24	0.22765	-3.031	0	0	23			
25	-0.8356	-1.7196	0	0	24			
26	-0.4651	-1.7291	0	0	25			
27	0.99636	1.11536	0	0	26			
28	-0.4863	-0.6857	0	0	27			
29	0.32605	-0.6553	0	0	28			
30	0.06688	-1.795	0	0	29			
31	0.60765	-1.2828	0	0	30			
32	-0.5227	-1.0369	0	0	31			
33	0.00674	-0.4212	0	0	32			
34	-0.3698	-0.5022	0	0	33			
35	-0.0276	0.02405	0	0	34			
36	1.29679	1.89861	0	0	35			
37	0.64602	0.80529	0	0	36			
38	-0.4793	1.35729	0	0	37			
39	0.60355	-0.0246	0	0	38			
40	1.19061	-0.116	0	0	39			
41	0.70038	-1.0653	0	0	40			
42	0.42229	0.27986	0	0	41			
43	0.01915	-0.3428	0	0	42			
44	-0.9977	-0.4687	0	0	43			
45	0.62008	-0.1473	0	0	44			
46	0.5383	0.16788	0	0	45			
47	-1.3162	0.08772	0	0	46			
48	-1.8875	-1.4128	0	0	47			
49	-1.3405	0.74083	0	0	48			
50	1.00145	-0.0255	0	0	49			
51	0.42807	-0.0183	0	0	50			
52	POSX50	POSY50						

Figure 10.2
Deming Funnel Experiment: Rule 2

	A	B	C	D	E	F	G	H
1	DROPX	DROPY	POSX	POSY	**Deming Rule 2**		**Figure 10.2**	
2	0.85979	0.38508	0	0	1			
3	0.31351	0.51422	-0.8598	-0.3851	2		/DROPX	/DROPY
4	-2.0834	-2.2305	-1.1733	-0.8993	3		Output	Output
5	0.82833	2.05089	0.91014	1.33117	4		A51	B51
6	-0.0979	-0.7019	0.08181	-0.7197	5	MINIMUM	-4.543677	-3.84423
7	1.26273	1.74566	0.17968	-0.0178	6	MAXIMUM	3.791375	4.67255
8	0.02196	-2.0747	-1.0831	-1.7635	7	MEAN	0.001030	-0.00024
9	0.10956	1.3312	-1.105	0.31124	8	STD DEV	1.360722	1.50712
10	-0.725	-0.6451	-1.2146	-1.02	9	VARIANCE	1.851563	2.2714
11	-2.5302	0.11661	-0.4896	-0.3749	10			
12	0.94632	0.4953	2.04057	-0.4915	11			
13	1.36504	-1.0475	1.09425	-0.9868	12			
14	-0.8904	0.4219	-0.2708	0.06072	13			
15	-0.3915	-0.9833	0.61961	-0.3612	14			
16	1.46777	0.21506	1.01115	0.62208	15			
17	-0.1328	-0.1845	-0.4566	0.40702	16			
18	0.57471	0.46433	-0.3239	0.59152	17			
19	-0.6251	1.02611	-0.8986	0.1272	18			
20	-2.7515	0.01725	-0.2735	-0.8989	19			
21	2.49064	-0.9625	2.47805	-0.9162	20			
22	0.6974	-1.1955	-0.0126	0.04632	21			
23	-2.4933	0.99044	-0.71	1.24186	22			
24	1.5377	1.73745	1.78334	0.25142	23			
25	-0.6471	0.26555	0.24564	-1.486	24			
26	1.35057	-0.937	0.8927	-1.7516	25			
27	-1.2051	-2.0779	-0.4579	-0.8145	26			
28	-0.0271	1.19898	0.74718	1.26336	27			
29	0.68388	2.09398	0.77431	0.06439	28			
30	0.16916	-0.9028	0.09043	-2.0296	29			
31	1.40992	-0.9768	-0.0787	-1.1268	30			
32	-2.202	1.56048	-1.4887	-0.15	31			
33	0.93218	-0.9065	0.71338	-1.7105	32			
34	-0.7692	0.84414	-0.2188	-0.804	33			
35	1.58451	-2.4761	0.55044	-1.6481	34			
36	-1.229	-0.602	-1.0341	0.82802	35			
37	-0.1912	1.39605	0.19497	1.43	36			
38	0.44018	0.96536	0.38615	0.03395	37			
39	0.76764	-0.189	-0.054	-0.9314	38			
40	-0.7519	-0.5447	-0.8217	-0.7424	39			
41	-0.197	-0.2224	-0.0697	-0.1977	40			
42	0.5696	-2.2189	0.12724	0.0247	41			
43	-1.2267	1.44117	-0.4424	2.24362	42			
44	1.50513	1.21551	0.78432	0.80245	43			
45	-2.1721	-2.248	-0.7208	-0.4131	44			
46	2.06109	1.19263	1.45129	1.83492	45			
47	1.35303	1.62437	-0.6098	0.64229	46			
48	-1.0917	-0.9421	-1.9628	-0.9821	47			
49	-0.9358	-0.262	-0.8711	-0.04	48			
50	-0.3024	0.0014	0.06468	0.22206	49			
51	-1.4732	0.56368	0.36708	0.22066	50			
52	POSX50	POSY50						

Step 3 In D3 we compute the new y-coordinate for the funnel by entering the formula

```
Excel: =D2-B2

Lotus: +D2-B2
```

Step 4 In A3 we generate the x-coordinate for the second ball dropped by entering the formula

```
Excel: =C3+RISKNORMAL(0,1) or =RISKNORMAL(C3,1)

Lotus: +C3+@RISKNORMAL(0,1) or @RISKNORMAL(C3,1)
```

This ensures the x-coordinate for the second drop will on average equal the current funnel x-coordinate.

Step 5 Similarly, in cell B3 we generate the y-coordinate for the second funnel drop by entering the formula

```
Excel: =D3+RISKNORMAL(0,1) or =RISKNORMAL(D3,1)

Lotus: +D3+@RISKNORMAL(0,1) or @RISKNORMAL(D3,1)
```

Step 6 Copying from the range A3:D3 to A3:D51yields the x- and y-coordinates for all 50 drops.

Using @RISK we simulated the x-coordinate for the 50th drop 900 times. We found an average x-coordinate of .001 and a variance of 1.85. We found an average y-coordinate of –.0002 and a variance of 2.27. Thus Rule 2 appears to double (as compared to Rule 1) the variance of each coordinate of the 50th drop!

10.3 Comparison of Rules 1–4

In Problems 10.1 and 10.2 we ask you to simulate Rule 3 and Rule 4. For each rule, Table 10.1 summarizes the results of a simulation of the x-coordinate of the 50th ball.

Table 10.1 Summary of Funnel Experiment Results (50th Drop)

Rule #	Mean x-Coordinate	Variance of x-Coordinate
1	.0006(0)	.995 (1)
2	.001(0)	1.85 (2)
3	+.0001(0)	51.9! (50)
4	–.0002(0)	49.77! (50)

The theoretical mean and variance of the x-coordinate on the 50th drop are given in parentheses in Table 10.1. See Section 10.5 for a derivation of these results. Table 10.1 makes it clear that not touching the funnel is by far the best strategy. Any attempts to tamper with or adjust a stable process are doomed to failure! By the way, for Rule 3 and Rule 4 the variance of the x-coordinate on the *n*th drop equals *n*. This means that the variance of the x-coordinate continually increases!!

10.4 Lesson of the Funnel Experiment

The moral of the funnel experiment is "If it ain't broke, don't fix it." That is, adjusting a stable process will make things much worse! All systems have a certain level of inherent variability. By keeping control charts, we can greatly decrease the chances that we will increase the variability of a system by tampering with the system when it is stable (a stable system is often called an **in-control system).**

10.5 Mathematical Explanation of the Funnel Experiment

We assume that the x- or y-coordinate of a ball is given by

$$\text{x-Coordinate of Ball} = \text{Funnel x-Coordinate} + \text{Error Term } \epsilon$$

where ϵ is normal with mean 0 and standard deviation 1

Remember that

> For independent random variables the variance of the sum of random variables equals the sum of the variances of the individual random variables.
>
> Variance ϵ_i = Variance $-\epsilon_i$.
>
> ϵ_i = Error Term for Drop *i* x-coordinate.

Table 10.2 Rule 1 Don't Touch Funnel

Drop #	Funnel x-Coordinate	Ball x-Coordinate	Variance of Ball x-Coordinate
1	0	ϵ_1	1
2	0	ϵ_2	1
3	0	ϵ_3	1
4	0	ϵ_4	1

Thus, on any drop, the variance of the x-coordinate of the ball is 1.

Table 10.3 Rule 2 Adjust Relative to Last Position

Drop #	Funnel x-Coordinate	Ball x-Coordinate	Variance of Ball x-Coordinate
1	0	ϵ_1	1
2	$-\epsilon_1$	$\epsilon_2-\epsilon_1$	2
3	$-\epsilon_2=-\epsilon_1-(\epsilon_2-\epsilon_1)$	$\epsilon_3-\epsilon_2$	2
4	$-\epsilon_3=-\epsilon_2-(\epsilon_3-\epsilon_2)$	$\epsilon_4-\epsilon_3$	2

Thus, after the first drop, the variance of the ball's x-coordinate is 2 (double Rule 1!).

Table 10.4 Rule 3 Adjust Funnel Relative to Target

Drop #	Funnel x-Coordinate	Ball x-Coordinate	Variance of Ball x-Coordinate
1	0	ϵ_1	1
2	$-\epsilon_1$	$\epsilon_2-\epsilon_1$	2
3	$-(\epsilon_2-\epsilon_1)$	$\epsilon_1-\epsilon_2+\epsilon_3$	3
4	$-(\epsilon_1-\epsilon_2+\epsilon_3)$	$-\epsilon_1+\epsilon_2-\epsilon_3+\epsilon_4$	4

Thus the variance of the ball's x-coordinate on Drop $n = n$! After a lot of drops we are clearly in trouble.

Table 10.5 Rule 4 Set Funnel Over Last Drop

Drop #	Funnel x-Coordinate	Ball x-Coordinate	Variance of Ball x-Coordinate
1	0	ϵ_1	1
2	ϵ_1	$\epsilon_1+\epsilon_2$	2
3	$\epsilon_1+\epsilon_2$	$\epsilon_1+\epsilon_2+\epsilon_3$	3
4	$\epsilon_1+\epsilon_2+\epsilon_3$	$\epsilon_1+\epsilon_2+\epsilon_3+\epsilon_4$	4

Thus, on Drop n, the variance of the ball's x-coordinate is n.
After a lot of drops we are in trouble.

Problems

Group A

10.1 Simulate 900 trials of the 50th drop for Rule 3. Then evaluate the average and variance of the x-coordinate for the 50th drop.

10.2 Simulate 900 trials of the 50th drop for Rule 4. Then evaluate the average and variance of the x-coordinate for the 50th drop.

10.3 A car company paints all of its cars yellow. In the company vault is a sample of the original yellow used to paint this year's cars. Suppose that each time a new batch of yellow paint is mixed, the company matches the new batch to the most recent batch (not the original). Which rule does this illustrate?

10.4 Suppose that missile tests indicate that on the average, a missile misses the target by 50 feet. Suppose a missile is fired and lands 80 feet east of the target. Then the army adjusts the control mechanism on the missile to a setting 80 feet west of the current setting. Which rule is being followed here?

Reference

Deming, E. 1986. *Out of the Crisis*. Cambridge: MIT Center for Advanced Engineering Study.

CHAPTER 11

the Taguchi Loss Function

Consider a color TV in which color density is considered perfect if it measures exactly 100. Suppose that manufacturers have set specifications that assert that a TV possesses satisfactory color density if its color density is between 95 and 105. If a color TV has a color density of 95 or 105, a repair costing $4 must be made. Japanese TV manufacturers find that the color density on their TVs is normally distributed with mean 100 and standard deviation 5/3. American TV manufacturers find that the color density on their TVs is equally likely to be any number between 95 and 105.

The probability that a Japanese TV will fail to meet specifications is the probability that a normal random variable will be more than (105 – 100)/(5/3) = 3 standard deviations away from its mean. This is around .003. An American TV will *always* meet specifications. For this reason, Americans claim their TVs are of better quality. Unfortunately, the Japanese TVs have a much better reliability record. Why?

The answer to this question lies in the fact that whenever a TV fails to have a color density of exactly 100, the customer suffers a quality loss. The further away the TV's color density is from 100, the larger the quality loss. The Japanese and American TVs both average out to a color density of 100, but the Japanese TVs exhibit much less variability. Thus the Japanese TVs are much more likely to have a color density near 100. This phenomenon explains why they perform better!

The Japanese quality expert Genichi Taguchi invented the **Taguchi Loss Function** to enable a firm to quantify its loss from poor product quality. Let T = the target measurement for the product. For color TV color density we have $T = 100$. If a product has a measurement x, Taguchi postulates that the loss due to poor product quality is given by

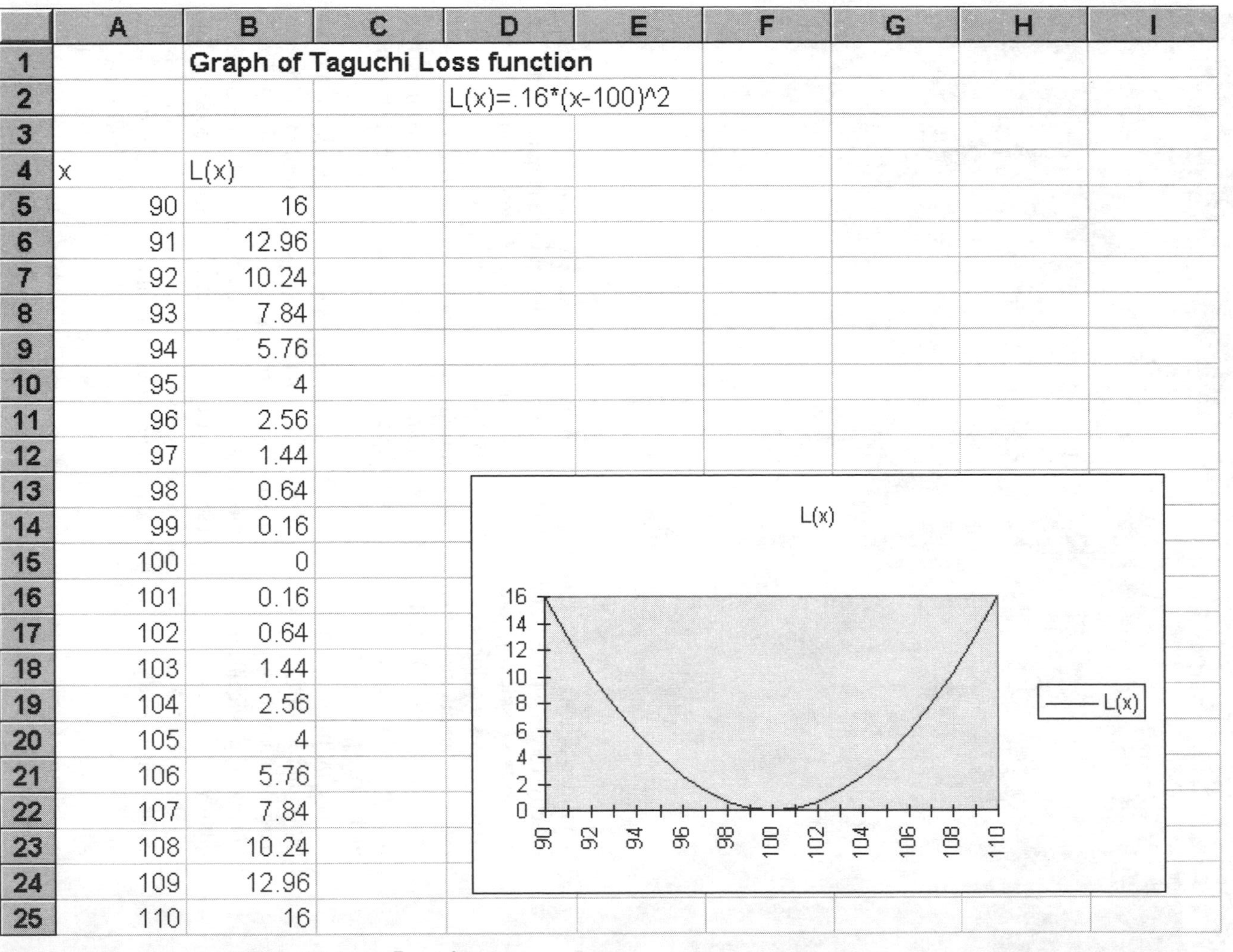

	A	B	C	D	E
1		**Graph of Taguchi Loss function**			
2				L(x)=.16*(x-100)^2	
3					
4	x	L(x)			
5	90	16			
6	91	12.96			
7	92	10.24			
8	93	7.84			
9	94	5.76			
10	95	4			
11	96	2.56			
12	97	1.44			
13	98	0.64			
14	99	0.16			
15	100	0			
16	101	0.16			
17	102	0.64			
18	103	1.44			
19	104	2.56			
20	105	4			
21	106	5.76			
22	107	7.84			
23	108	10.24			
24	109	12.96			
25	110	16			

Figure 11.1 Graph of Taguchi Loss Function

$$L(x) = k(x - T)^2 \quad \textbf{(11.1)}$$

$L(x)$ is called the Taguchi Loss Function. $L(x)$ equals the expected quality loss incurred by the company if a product with measurement x is produced. In (11.1) k is used to "calibrate" the loss function. For example, we know that $L(105) = \$4$. Thus

$$\$4 = k(105 - 100)^2 \text{ or } 25k = \$4 \text{ or } k = \$.16$$

The loss function $L(x) = .16*(x - 100)^2$ is graphed in Figure 11.1. Notice the loss due to poor quality increases at a larger rate the farther we move away from the target.

11.1 Using@RISK to Quantify Quality Loss

We can now use @RISK to compare the quality loss from Japanese and American TVs. Our work is in the file Taguch.xls or Taguch.wk4. Our spreadsheet is displayed in Figure 11.2.

We proceed as follows:

Step 1 Since we are going to use range names, we entered labels for the target, k, mean of Japanese TVs, standard deviation for Japanese TVs, color density for Japanese TVs in the cell range A3:A7.

Step 2 Select the range A3:B7 and choose Insert Name Create. Finally, tell Excel to create range names in the left column. This assigns, for example, the range name Target to cell B3.

Step 3 Enter the problem parameters (including the mean and standard deviation for Japanese TVs) in the cell range B3:B6.

Step 4 Since the color density for Japanese TVs follows a normal distribution, we generate the color density for a typical Japanese TV by entering in cell B7 the formula

```
Excel:  =RISKNORMAL(Mean,Std_Dev)

Lotus:  @RISKNORMAL(Mean,Std_Dev)
```

Observe that a range name must be one word!

Step 5 We compute the quality loss for a typical Japanese TV in cell B8 with the formula

```
Excel:  =k*(Color_Density-Target)^2

Lotus:  +k*(Color_Density-Target)^2
```

	A	B	C	D	E	F
1	**The Taguchi Loss Function**	**Figure 11.2**				
2	Japanese TV			US TV		
3	Target	100		Color Density	104.199	
4	k	0.16		Loss US	2.820989	
5	Mean	100				
6	Std Dev	1.666667				
7	Color Density	100.866				
8	Loss Japanese	0.119995				
9						
10		Simulation Results for TAGUCH.XLS				
11						
12		Iterations= 400				
13		Simulations= 1				
14		# Input Variables= 2				
15		# Output Variables= 2				
16		Sampling Type= Latin Hypercube				
17		Runtime= 00:00:21				
18						
19		Summary Statistics				
20						
21		Cell	Name	Minimum	Mean	Maximum
22						
23		B8	Loss Japa	1.39E-07	0.444984	4.767154
24		E4	Loss US	9.53E-08	1.334138	3.989971

Figure 11.2 Taguchi Loss TV Example

Step 6 Since the color density for an American TV follows a uniform distribution, we generate the color density for a typical American TV by entering in cell E3 the formula

```
Excel: =RISKUNIFORM(95,105)

Lotus: @RISKUNIFORM(95,105)
```

Step 7 In cell E4 we compute the quality loss for a typical American TV with the formula

```
Excel: =k*(E3-Target)^2

Lotus: +k*(E3-Target)^2
```

Step 8 We select cells B8 and E4 as output cells and run 400 iterations. We find the average quality loss per US TV ($1.33) to be triple the average quality loss per Japanese TV ($.44)!

Remarks

1 For some products, the target is a measurement of 0, and a quality loss is associated with a larger measurement. This is known as the "smaller-is-better" case. In this situation, the proper loss function is $L(x) = kx^2$.

2 For some products, the larger the measurement the higher the quality. This is the "larger-is-better" case. In this situation, the proper loss function is $L(x) = k/x^2$.

3 For further discussion of the Taguchi Loss Function, we refer the reader to Ealey (1988).

Problems

Group A

11.1 Keefer Paper produces pulp that is used for paper recycling. For a particular grade of paper the target level of brightness is 80. If the brightness level of paper is 85, there is a 10% chance that a customer will complain, and a cost of $50 per ton will be incurred due to warranty costs and lost goodwill. Keefer's present machines produce paper that is equally likely to have a brightness between 72 and 88. The cost of running this machine is $200,000 per year.

A new machine would cost $400,000 per year to operate. It would produce paper with a brightness level that follows a normal distribution with mean 80 and standard deviation of 1.

Keefer produces 60,000 tons of paper per year. Can they justify the use of the new machine on the grounds of quality improvement?

11.2 The target value for a piece of automobile weatherstripping is 20mm. If the weatherstripping is 15mm, half of all consumers will request a $50 repair. The mean and standard deviation of the weatherstripping from three suppliers is given in Table 11.1.

Table 11.1

Supplier	Mean	Standard Deviation
1	20	5/3
2	18	1
3	17	3/5

From which supplier would you purchase weatherstripping? Assume all measurements are normally distributed.

11.3 Consider a weld that must be on-center. If the weld is more than 1mm off-center repairs costing $100 must be made. Our welds have averaged .25mm off-center with a standard deviation of .10mm. Estimate the average quality loss per weld. Assume all measurements are normally distributed.

Reference

Ealey, L. 1988. *Quality by Design*. Dearborn, MI: American Supplier Institute.

CHAPTER 12

The Use of Simulation in Project Management

Consider a project made up of many activities. For each activity, there is a set of activities (called the **predecessors of the activity**) that must be completed before the activity begins. Assuming that the duration of each activity is known with certainty, it is easy (see Section 8.4 of Winston 1994) to determine how long the project will take to complete and to identify **critical activities**. An activity is critical if increasing the duration of the activity by a small amount Δ will increase the length of time needed to complete the project by Δ. A simple example of such a project follows.

Example 12.1 The Widgetco Example

Widgetco is about to introduce a new product (Product 3). One unit of Product 3 is produced by assembling 1 unit of Product 1 and 1 unit of Product 2. Before production begins on either Product 1 or Product 2, raw materials must be purchased and workers must be trained. Before Products 1 and 2 can be assembled into Product 3, the finished Product 2 must be inspected. A list of activities and their predecessors and the duration of each activity is given in Table 12.1.

Table 12.1

Activity	Predecessors	Duration (days)
A = train workers	–	6
B = purchase raw materials	–	9
C = produce Product 1	A, B	8
D = produce Product 2	A, B	7
E = test Product 2	D	10
F = assemble Products 1 and 2	C, E	12

The predecessor relationships given in Table 12.1 can be expressed in a project diagram. The project diagram for this example is given in Figure 12.1. Node 4 indicates, for example, that Activity F cannot begin until Activities C and E are complete. It is easy to show that this project will take 38 days to complete and that activities B, D, E, and F are critical. This implies that if the duration of any of these activities increases, then the time needed to complete the project will also increase.

Figure 12.1
Project Diagram

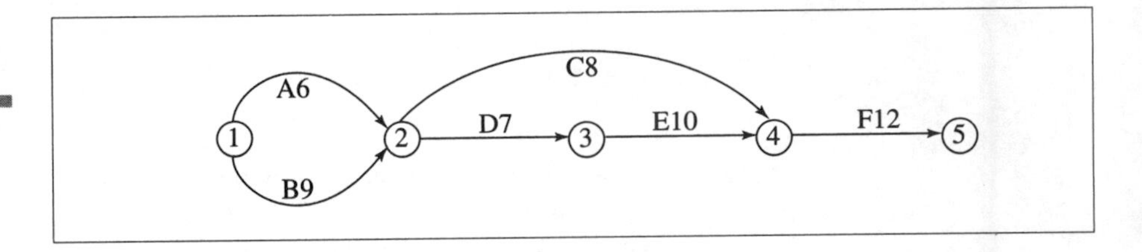

Our prior discussion requires the assumption that the duration of each activity is known with certainty. In reality, the time required to complete each activity is uncertain. Of course, this implies that the length of time needed to complete the project will also be uncertain. The fact that the duration of each project is uncertain also implies that *for any activity there is a probability (not necessarily equal to 0 or 1) that the activity is critical.*

To illustrate the previous discussion, suppose that for the Widgetco example the duration of Activities A–F follow independent normal distributions with the parameters given in Table 12.2.

Table 12.2 Duration of Activities in Widgetco Example

Activity	Mean	Standard Deviation
A	6	2
B	9	1
C	8	2
D	7	2
E	10	3
F	12	3

If the duration of each activity is known with certainty, B is a critical activity. This follows because B took more time than A. If the duration of each activity is random as described in Table 12.2, it is now possible, however, for A to take longer than B. Then A would be a critical activity! In this section we will learn how to use simulation to estimate the probability that an activity is critical.

From the data in Table 12.2 it is also apparent that the project may take more (or less) time than 38 days to complete. We now show how to use simulation to estimate the probability distribution of the length of time needed to complete the project.

12.1 Estimating Probability Distribution of Project Completion Time

In Figure 12.2 (file Pert.wk4 or Pert.xls) we give the spreadsheet needed to estimate the probability distribution for project completion time for the project described by Table 12.2 and Figure 12.1. We proceed as follows:

Step 1 In cells B3:G3 we generate the duration of each activity by entering the following formulas (see Figure 12.2a):

```
Excel

B3:  =RISKTNORMAL(6,2,0,16)
C3:  =RISKTNORMAL(9,1,0,14)
D3:  =RISKTNORMAL(8,2,0,18)
E3:  =RISKTNORMAL(7,2,0,17)
F3:  =RISKTNORMAL(10,3,0,25)
G3:  =RISKTNORMAL(12,3,0,27)

Lotus

B3:  @RISKTNORMAL(6,2,0,16)
C3:  @RISKTNORMAL(9,1,0,14)
D3:  @RISKTNORMAL(8,2,0,18)
E3:  @RISKTNORMAL(7,2,0,17)
F3:  @RISKTNORMAL(10,3,0,25)
G3:  @RISKTNORMAL(12,3,0,27)
```

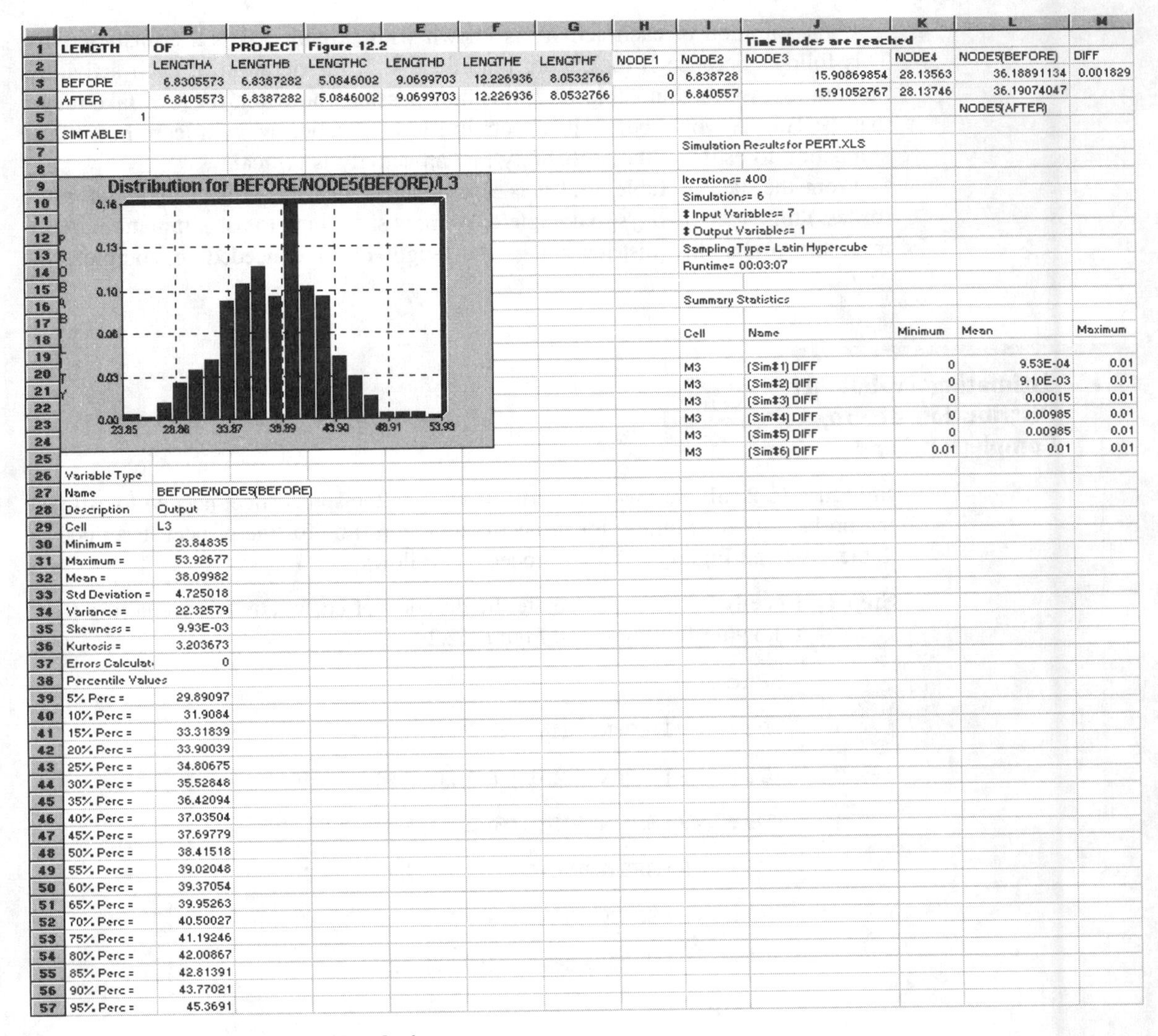

	A	B	C	D	E	F	G	H	I	J	K	L	M
1	LENGTH	OF	PROJECT	Figure 12.2						Time Nodes are reached			
2		LENGTHA	LENGTHB	LENGTHC	LENGTHD	LENGTHE	LENGTHF	NODE1	NODE2	NODE3	NODE4	NODE5(BEFORE)	DIFF
3	BEFORE	6.8305573	6.8387282	5.0846002	9.0699703	12.226936	8.0532766	0	6.838728	15.90869854	28.13563	36.18891134	0.001829
4	AFTER	6.8405573	6.8387282	5.0846002	9.0699703	12.226936	8.0532766	0	6.840557	15.91052767	28.13746	36.19074047	
5	1											NODE5(AFTER)	
6	SIMTABLE!												

Simulation Results for PERT.XLS

Iterations= 400
Simulations= 6
# Input Variables= 7
# Output Variables= 1
Sampling Type= Latin Hypercube
Runtime= 00:03:07

Summary Statistics

Cell	Name	Minimum	Mean	Maximum
M3	(Sim#1) DIFF	0	9.53E-04	0.01
M3	(Sim#2) DIFF	0	9.10E-03	0.01
M3	(Sim#3) DIFF	0	0.00015	0.01
M3	(Sim#4) DIFF	0	0.00985	0.01
M3	(Sim#5) DIFF	0	0.00985	0.01
M3	(Sim#6) DIFF	0.01	0.01	0.01

	A	B
26	Variable Type	
27	Name	BEFORE/NODE5(BEFORE)
28	Description	Output
29	Cell	L3
30	Minimum =	23.84835
31	Maximum =	53.92677
32	Mean =	38.09982
33	Std Deviation =	4.725018
34	Variance =	22.32579
35	Skewness =	9.93E-03
36	Kurtosis =	3.203673
37	Errors Calculat	0
38	Percentile Values	
39	5% Perc =	29.89097
40	10% Perc =	31.9084
41	15% Perc =	33.31839
42	20% Perc =	33.90039
43	25% Perc =	34.80675
44	30% Perc =	35.52848
45	35% Perc =	36.42094
46	40% Perc =	37.03504
47	45% Perc =	37.69779
48	50% Perc =	38.41518
49	55% Perc =	39.02048
50	60% Perc =	39.37054
51	65% Perc =	39.95263
52	70% Perc =	40.50027
53	75% Perc =	41.19246
54	80% Perc =	42.00867
55	85% Perc =	42.81391
56	90% Perc =	43.77021
57	95% Perc =	45.3691

Figure 12.2 Critical Path Simulation

For instance, the formula in cell B3 ensures that we generate the duration of Activity A by sampling from a normal distribution with mean 6 and standard deviation 2 and ignoring any generated observations that yield a negative duration for the activity.

Step 2 We can now express (in terms of the durations of the activities) the time that each node in the network is reached. In H3 we enter 0 to indicate that Node 1 (representing start of the project) is reached at Time 0.

	A	B	C	D	E	F	G	H	I	J	K	L	M
1	LENGTH	OF	PROJECT	Figure 12.2									
2		LENGTHA	LENGTHB	LENGTHC	LENGTHD	LENGTHE	LENGTHF	NODE1	NODE2	Time Nodes are reached NODE3	NODE4	NODE5(BEFORE)	DIFF
3	BEFORE	6.83055733	6.83872819	5.08460023	9.06997034	12.2269362	8.05327664	0	6.838728	15.90869854	28.13563	36.18891134	0.001829
4	AFTER	6.84055733	6.83872819	5.08460023	9.06997034	12.2269362	8.05327664	0	6.840557	15.91052767	28.13746	36.19074047	
5	1												
6	SIMTABLE!											NODE5(AFTER)	

Figure 12.2a

Step 3 In I3 we enter

```
Excel:  =MAX(B3,C3)
Lotus:  @MAX(B3,C3)
```

to indicate that Node 2 is reached when both A and B are completed.

Step 4 In J3 we enter

```
Excel:  =E3+I3
Lotus:  +E3+I3
```

to indicate that Node 3 is reached at
(Time Node 2 Is Reached) + (Duration of D)

Step 5 In K3 we enter

```
Excel:  =MAX(I3+D3,J3+F3)
Lotus:  @MAX(I3+D3,J3+F3)
```

to indicate that Node 4 occurs at

max} (Time Node 2 Occurs) + (Duration of C)
(Time Node 3 Occurs) + (Duration of E)

Step 6 Finally, in cell L3 we enter

```
Excel:  =K3+G3
Lotus:  +K3+G3
```

to indicate that Node 5 (completion of the project) occurs at

(Time Node 4 Occurs) + (Duration of F)

We now used @RISK to simulate the length of the project 400 times. The output cell was L3. A histogram of the project length is shown in Figure 12.2b. From the statistics portion of the @RISK results we found the percentiles of project length given in Table 12.3.

Figure 12.2b

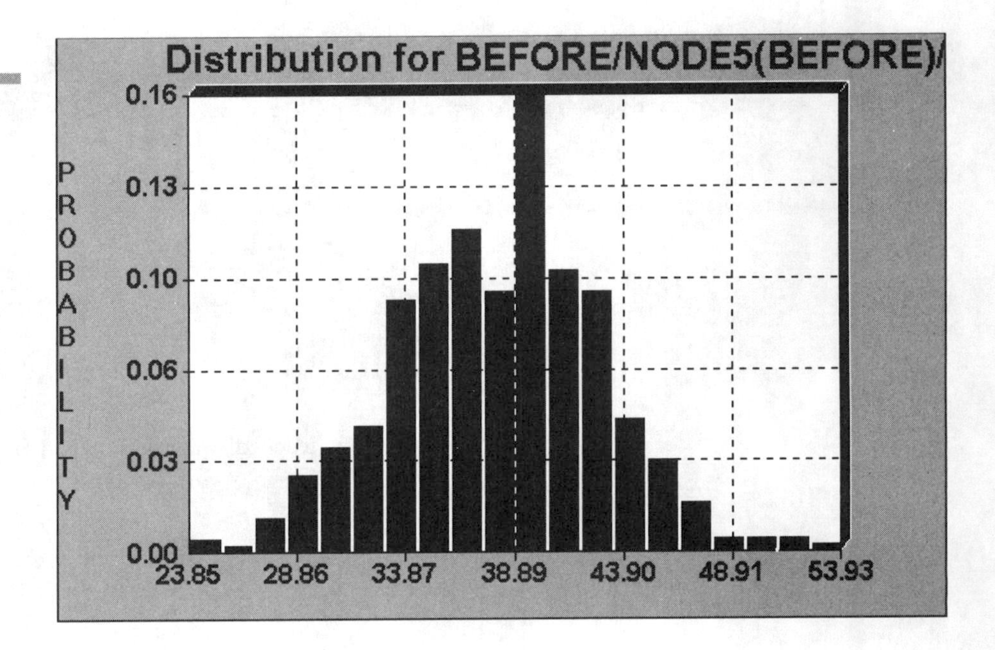

Table 12.3 Percentiles of Project Completion Time

Percentile	Project Length
10th	31.91
20th	33.90
30th	35.53
40th	37.04
50th	38.42
60th	39.37
70th	40.50
80th	42.01
90th	43.77
95th	45.37

For example, from Table 12.3 we find that our simulation indicates that there is a 5% chance the project's length will exceed 45.37 days and there is a 20% chance that the project will take 33.90 days or less (see Figure 12.2c).

Figure 12.2c

Variable Type			
Name	BEFORE/NODE5(BEFORE)		
Description	Output		
Cell	L3		
Minimum =	23.84835		
Maximum =	53.92677		
Mean =	38.09982		
Std Deviation =	4.725018		
Variance =	22.32579		
Skewness =	9.93E-03		
Kurtosis =	3.203673		
Errors Calculat	0		
Percentile Values			
5% Perc =	29.89097		
10% Perc =	31.9084		
15% Perc =	33.31839		
20% Perc =	33.90039		
25% Perc =	34.80675		
30% Perc =	35.52848		
35% Perc =	36.42094		
40% Perc =	37.03504		
45% Perc =	37.69779		
50% Perc =	38.41518		
55% Perc =	39.02048		
60% Perc =	39.37054		
65% Perc =	39.95263		
70% Perc =	40.50027		
75% Perc =	41.19246		
80% Perc =	42.00867		
85% Perc =	42.81391		
90% Perc =	43.77021		
95% Perc =	45.3691		

12.2 Determining the Probability That an Activity Is Critical

Recall that an activity is critical if increasing the activity's length by a small amount Δ increases the length of the project by the same amount Δ. We can use simulation to estimate the probability that an activity is a critical activity. We proceed as follows:

Step 1 Enter in cell A5 the formula

```
Excel: =RISKSIMTABLE({1,2,3,4,5,6})

Lotus: @RISKSIMTABLE(1,2,3,4,5,6,)
```

This formula will allow us to run six simulations. The results of Simulation *i* will be used to determine the probability that the activity corresponding to the *i*th letter of the alphabet is critical.

Step 2 In cells B4:G4 we enter the following formulas:

```
Excel
B4:  =B3+IF(A5=1,.01,0)
C4:  =C3+IF(A5=2,.01,0)
D4:  =D3+IF(A5=3,.01,0)
E4:  =E3+IF(A5=4,.01,0)
F4:  =F3+IF(A5=5,.01,0)
G4:  =G3+IF(A5=6,.01,0)
Lotus
B4:  +B3+@IF(A5=1,.01,0)
C4:  +C3+@IF(A5=2,.01,0)
D4:  +D3+@IF(A5=3,.01,0)
E4:  +E3+@IF(A5=4,.01,0)
F4:  +F3+@IF(A5=5,.01,0)
G4:  +G3+@IF(A5=6,.01,0)
```

During simulation *i*, cell A5 will assume the value *i*. Then cells B4:G4 will assume the durations of activities A–F that were generated in row 3, *except for the fact that the duration of activity* i *will be increased by .01.*

Step 3 We now copy the formulas in the range H3:L3 to the range H4:L4. The entry in cell L4 during Simulation *i* will now equal the length of the project when the duration of Activity *i* is increased by .01. Now enter into cell M3 the formula

```
Excel:  =L4-L3
Lotus:  +L4-L3
```

In each iteration of any simulation the value assumed by cell M3 will equal 0 or .01. We can estimate the *probability that activity* i *is critical by the fraction of the Simulation* i *iterations that cause cell M3 to equal .01!* In Figure 12.2d we see the results for cell M3 for each simulation.

Cell	Name	Minimum	Mean	Maximum
M3	(Sim#1) DIFF	0	9.53E-04	0.01
M3	(Sim#2) DIFF	0	9.10E-03	0.01
M3	(Sim#3) DIFF	0	0.00015	0.01
M3	(Sim#4) DIFF	0	0.00985	0.01
M3	(Sim#5) DIFF	0	0.00985	0.01
M3	(Sim#6) DIFF	0.01	0.01	0.01

Figure 12.2d

For example, in Simulation 1 the duration of Activity A was increased by .01, and the mean of cell M3 was .00095. Let p = fraction of Simulation 1 iterations in which cell M3 assumes the value .01. Then $p * (.01) + (1 - p) * 0 = .00095$. This implies that $p = .00095/.01 = .095$. Thus we estimate that there is a 9.5% chance that Activity A will be a critical activity. Our estimates of the probability that each activity is critical are given in Table 12.4 (see also Figure 12.2d).

Table 12.4 Estimation of Probability That Activities Are Critical

Activity	Probability of Being Critical
A	.095
B	.910
C	.015
D	.985
E	.985
F	1.000

12.3 The Beta Distribution and Project Management

In closing, we would like to note that the duration of an activity is usually not modeled by a normal distribution, but by a **beta distribution**. A beta distribution is a skewed distribution (unlike the symmetric normal distribution) that allows the duration of an activity to occasionally assume very large values but to rarely assume very small values. @RISK enables you to simulate a beta distribution. We now explain how to use the beta distribution to model the duration of an activity in a project management simulation.

	A	B	C	D	E
1	Simulating a beta distribution				Figure 12.3
2	NORMMEAN	NORMST	MAX	MEAN	STDEV
3	0.428571429	0.142857	14	6	2
4	A1	A2			
5	4.714285714	6.285714	simulated		
6			beta		
7			6.731411		
8	MEAN				
9	0.428571429				
10	STDDEV				
11	0.142857143				

Figure 12.3 Estimating Parameters of a Beta Distribution

The beta distribution included in @RISK assumes values between 0 and 1 and is determined by two parameters, α_1 and α_2. Suppose we are given for an activity its mean duration, maximum duration, and standard deviation of its duration. How can we use @RISK to simulate the duration of the activity? Simply input the maximum (call it Max), mean, and standard deviation of the activity's duration in the cell range C3:E3 of Figure 12.3 (file Beta.wk4 or Beta.xls). Then the values of α_1 and α_2 that yield the correct mean and standard deviation of the activity are computed in the cell range A5:B5. Finally, you would simulate the duration of the activity with the formula

```
Excel:  =MAX*RISKBETA(α1,α2)
Lotus:  MAX*@RISKBETA(α1,α2)
```

To illustrate, consider an activity that has a mean duration of 6 days, a standard deviation of 2 days, and a maximum duration of 14 days. From Figure 12.3 we find $\alpha_1 = 4.71$ and $\alpha_2 = 6.29$. We could then generate the duration of the activity with the formula

```
Excel:  =14*RISKBETA(4.71,6.29)
Lotus:  14*@RISKBETA(4.71,6.29)
```

Remarks

1. It may be easier to obtain estimates of an activity's most optimistic, most likely, and most pessimistic duration than it is to obtain estimates of the mean and standard deviation of an activity's duration. In this situation it is probably best to use the triangular distribution to model the duration of an activity.

2 Tornado diagrams can be used to examine which activities have the most influence on the project's completion time.

3 The beta distribution is very useful in modeling random variables that are nonsymmetric. For instance, if car sales for a year are nonsymmetric, then the beta distribution can be used to model annual car sales.

Problems

Group A

12.1 The city of Bloomington is about to build a new water treatment plant. Once the plant is designed (D) we can select the site (S), the building contractor (C) and the operating personnel (P). Once the site is selected we may erect the building (B). We can order the water treatment machine (W) and prepare the operations manual (M) only after the contractor is selected. We can begin training (T) the operators when both the operations manual and the operating personnel selection are completed. When the machine arrives and the building is finished, we can install the treatment machine (I). Once the treatment machine is installed and operators are trained, we can obtain our operating license (L). The estimated mean and standard deviation of the time (in months) needed to complete each activity is given in Table 12.5.

Table 12.5

Activity	Mean	Standard Deviation
D	6	1.5
S	2	0.3
C	4	1.0
P	3	1.0
B	24	6.0
W	14	4.0
M	3	0.4
T	4	1.0
I	6	1.0
L	3	0.6

a Use simulation to estimate the probability that the project will be completed in under 50 days.

b Estimate the probiability that the project will be completed in more than 55 days.

c Estimate the probability that B, I, and T are critical activities.

12.2 In order to complete an addition to the Business Building, the following activities need to be completed (all times are in months):

Table 12.6

Activity	Predecessors	Mean Time	Standard Deviation
A Hire workers	—	4	0.6
B Dig big hole	A	9	2.5
C Pour foundation	B	5	1.0
D Destroy Room 111	A	7	2.0
E Build main structure	C	10	1.5

The project is completed once Room 111 has been destroyed and the main structure has been built.

a Estimate the probability that it will take at least three years to complete the addition.

b Estimate the probability that each activity will be a critical activity.

12.3 In order to build Indiana University's new law building, the following activities must be completed (all times are in months):

Table 12.7

Activity	Predecessors	Mean Time	Standard Deviation
A Obtain funding	—	6	0.6
B Design building	A	8	1.3
C Prepare site	A	2	0.2
D Lay foundation	B, C	2	0.3
E Erect walls and roof	D	3	1.0
F Finish exterior	E	3	0.6
G Finish interior	D	7	1.5
H Landscape grounds	F, G	5	1.2

a Estimate the probability that the project will take under 30 months.

b Estimate the probability that the project will take more than three years.

c Estimate the probability that the following activities are critical: A, B, C, and G.

CHAPTER 13

Simulating Craps (and Other Games)

Using a spreadsheet to simulate the game of craps is an instructive exercise. It is an excellent example of the power of logical `IF` statements in simulation.

Example 13.1

The game of craps is played as follows: We throw two dice. If the total on the first throw is 2, 3, or 12, we lose the game. If the total on the first throw is 7 or 11 we win the game. If any other total occurs on the first throw we keep throwing the dice until we either throw a 7 or throw the same number we rolled on the first throw (our "point"). If we roll a 7 we lose. If we roll our point, we win. Use simulation to estimate the probability of winning at craps.

Solution From basic probability (just list all possible ways two dice can be tossed) we find that the probability mass function for the total given on any throw is as follows:

Table 13.1

Result of Throw	Probability of Result
2 or 12	1/36
3 or 11	2/36
4 or 10	3/36
5 or 9	4/36
6 or 8	5/36
7	6/36

	A	B	C	D	E	F	G	H	AD	AE
1	**CRAPS**	**SIMULATION**		**Figure 13.1**						
2	TOSS#	1	2	3	4	5	6	7	29	30
3	DIE TOSS	4	8	5	10	5	11	6	6	8
4	GAME STATUS	2	2	2	2	2	2	2	0	0
5	0=LOSS	WIN??	0							
6	1=WIN									
7	2=STILL GOING	95% CI		Simulation Results for CRAPS.XLS						
8	LOWER									
9	UPPER			Iterations= 900						
10				Simulations= 1						
11				# Input Variables= 30						
12				# Output Variables= 1						
13				Sampling Type= Latin Hypercube						
14				Runtime= 00:03:58						
15										
16				Summary Statistics						
17										
18				Cell	Name	Minimum	Mean	Maximum		
19										
20				C5	WIN??	0	0.476667	1		
21	95% CI for WInning at Craps									
22	Standard Deviation		0.499455							
23	Lower Limit		0.443992							
24	Upper Limit		0.509341							
25	Sample Size		900							

Figure 13.1 Craps Simulation

Thus, for example, a fraction 3/36 of all tosses yields a 4, 5/36 a 6, etc. Our simulation is in Figure 13.1 (file Craps.wk4 or Craps.xls). As we think about setting up our spreadsheet, we see a problem. Craps can theoretically go on forever. How many rolls of the dice should we allow before we end a game? Observe that the game is most likely to go for a long time if the first roll of the dice results in a point that is hard to make. The hardest point to make is 4 or 10. If we toss a 4 on the first roll, then on any later roll the probability that the game will end is 3/36 + 6/36 = 1/4, and the probability that the game will continue is 1 – 1/4 = 3/4. Thus if we play the game for 30 rolls the probability that the game will not be over cannot exceed $(3/4)^{29}$ = .0002. With this motivation, we will end each game after 30 rolls of the dice.

Step 1 In cell B3 we generate the result of the first roll of the dice by entering the formula

```
Excel: =RISKDISCRETE({2,3,4,5,6,7,8,9,10,11,12},
{1,2,3,4,5,6,5,4,3,2,1})
```

```
Lotus: @RISKDISCRETE(2,3,4,5,6,7,8,9,10,11,12,
1,2,3,4,5,6,5,4,3,2,1)
```

In cell B3 we are using a property of the `@RISKDISCRETE` or `=RISKDISCRETE` function that we have not previously discussed. If the "probabilities" assigned by the `@RISKDISCRETE` statement to each possible outcome do not add up to one, then @RISK treats them as weights and determines the probability of each outcome by dividing the outcome's weight by the sum of the weights for all possible outcomes. Since the weights for all outcomes add up to 36, this means that a 2 is assigned a probability of 1/36, a 3 a probability of 2/36, etc. Of course, this ensures that each possible roll of the dice occurs with the desired probability.

Copying the formula in cell B3 to the range C3:AE3 generates 29 more tosses of two dice.

Step 2 In row 4 we keep track of the result of the game. We use the following notation:

0 = Lost game

1 = Won game

2 = Game still going

The number in cell AE4 (copied to cell C5 so we can see it!) will indicate the final result of the game. In cell B4 we enter the formula

```
Excel: =IF(OR(OR((B3=2),(B3=3)),(B3=12)),0,
IF(OR((B3=7),(B3=11)),1,2))
```

```
Lotus: @IF((B3=2)#OR#(B3=3)#OR#(B3=12),
0,@IF(B3=7)#OR#(B3=11),1,2))
```

If the first toss is a 2, 3, or 12 this formula records a loss. If the first toss is 7 or 11, this formula records a win. For any other toss, this formula indicates that the game is not yet over.

Step 3 In cell C4 we enter the formula

```
Excel:  =IF(OR((B4=0),(B4=1)),B4,IF((C3-
$B3=0),1,IF((C3=7),0,2)))

Lotus:  @IF((B4=0)#OR#(B4=1),B4,@IF((C3-
$B3=0),1,@IF((C3=7),0,2)))
```

This formula enters the result of the first toss (win or loss) into cell C4. Thus if the game is over, no subsequent toss matters. If the game was not over on the first toss, then we check to see if the second toss matches the first toss. If so, we win! If this is not the case, we see if the second toss was a 7; if so, we lose. Otherwise the game goes on.

Copying this formula to the range D4:AE4 completes the simulation of one game of craps.

Step 4 We then select cell C5 as an output cell and run 900 iterations. We find that we won 47.7% of our games. Thus our simulation would estimate that we have a .477 chance of winning at craps. Using basic probability it can be shown that the actual probability of winning at craps is .493.

13.1 Confidence Interval for Winning at Craps

Recall from Chapter 3 that a 100(1 – α)% confidence interval for the expected value of a quantity p is given by

$$\bar{p} \pm t_{(\alpha/2,n-1)} s / n^{1/2} \qquad \textbf{(13.1)}$$

where $\bar{p}$ is the average value of p in the simulation and s is the standard deviation of p in the simulation. When a variable p can only assume the values 0 and 1, it can be shown that

$$s = \sqrt{\bar{p}(1-\bar{p})} = \sqrt{(0.477)(0.523)} = 0.4995$$

Applying formula (13.1) to our craps simulation we find a 95% confidence interval for the probability of winning at craps to be

$$.477-1.96*(.4995)/30 < \text{Probability of Winning} < .477-1.96*(.4995)/30$$

or

$$.444 < \text{Probability of Winning at Craps} < .509 \qquad \textbf{(13.2)}$$

See the cell range C22:C25 of Figure 13.1 for calculations of the 95% confidence interval for the probability of winning at craps.

The confidence interval formula in (13.1) assumes simple random sampling. Actually @RISK uses an improvement on simple random sampling, the **Latin hypercube method** (a version of stratified sampling). This ensures that the actual 95% confidence interval for winning at craps should be much narrower than indicated by (13.1). Unfortunately, the exact computation of a 95% confidence interval for the probability of winning at craps (with Latin hypercube sampling) is beyond the scope of this text.

In the problems the reader can try his or her hand at simulating many interesting games or investment situations. Have fun!!

Problems

Group A

13.1 The New York Knicks and Chicago Bulls are ready for the best of seven NBA Eastern finals. The two teams are evenly matched, but the home team wins 60% of the games between the two teams. The sequence of home and away games is to be chosen by the Knicks. The Knicks have the home edge and will be the home team for four of the seven scheduled games. They have the following choices (home team is listed for each game):

Sequence 1: NY, NY, CHIC, CHIC, NY, CHIC, NY

Sequence 2: NY, NY, CHIC, CHIC, CHIC, NY, NY

Use simulation to show that either sequence gives the Knicks the same chance of winning the series.

13.2 (Based on Kelly 1956) You currently have \$100. Each week you can invest any amount of money you currently have in a risky investment. With probability .4 the amount you invest is tripled (e.g., if you invest \$100, you increase your asset position by \$300), and with probability .6 the amount you invest is lost. Consider the following investment strategies:

1. Each week invest 10% of your money.
2. Each week invest 30% of your money.
3. Each week invest 50% of your money.

Simulate 100 weeks of each strategy 50 times. Which strategy appears to be best? In general, if you can multiply your investment by M with probability p and lose your investment with probability q, you should invest a fraction $\frac{p(M-1)-q}{M-1}$ of your money each week. This strategy maximizes (for a favorable game) the expected growth rate of your fortune and is known as the **Kelly Criteria.**

13.3 (Based on Marcus 1990) The Magellan mutual fund has beaten the Standard and Poor's 500 during 11 of the last 13 years. People use this as an argument that you can "beat the market." Here's another way to look at it that shows that Magellan's beating the market 11 out of 13 times is not unusual. Consider 50 mutual funds, each of which has a 50% chance of beating the market during a given year. Use simulation to estimate the probability that over a 13-year period the "best" of the 50 mutual funds will beat the market for at least 11 out of 13 years. This probability turns out to exceed 40%, which means that the best mutual fund's beating the market 11 out of 13 years is not an unusual occurrence!

13.4 You have made it to the final round of "Let's Make a Deal." You know that there is $1 million behind either Door 1, Door 2, or Door 3. It is equally likely that the prize is behind any of the three doors. The two doors without a prize have nothing behind them. You randomly choose Door 2. Before you see whether the prize is behind Door 2, Monty chooses to open a door which has no prize behind it. For the sake of definiteness, suppose that before Door 2 is opened Monty reveals that there is no prize behind Door 3. You now have the opportunity to switch and choose Door 1. Should you switch?

Use a spreadsheet to simulate this situation 400 times. For each "trial" use an @RISK function to generate the door behind which the prize lies. Then use another @RISK function to generate the door that Monty will open. Assume that Monty plays as follows: Monty knows where the prize is and will open an empty door, but he cannot open Door 2. If the prize is really behind Door 2, Monty is equally likely to open Door 1 or Door 3. If the prize is really behind Door 1, Monty must open Door 3. If the prize is really behind Door 3, Monty must open Door 1.

13.5 Star-crossed soap lovers Noah and Julia have had a big argument. Julia's sister Maria wants Noah and Julia to make up, so she has told them both to go to the romantic gazebo at 1 P.M. Unfortunately, Noah and Julia are not punctual. Each is equally likely to show up at the gazebo any time between 1 and 2 P.M. Assuming that each will stay for 20 minutes, what is the probability that they will meet?

13.6 The game of Chuck-a Luck is played as follows: You pick a number between 1 and 6 and toss three dice. If your number does not appear, you lose $1. If your number appears x times, you win $$x$. On the average, how much money will you win or lose on each play of the game?

13.7 I toss a die several times until the total number of spots I have seen is at least 13. What is the most likely total that will occur?

Group B

13.8 (Based on Morrison and Wheat 1984) When the team is behind late in the game, a hockey coach usually waits until there is one minute left before pulling the goalie. Actually coaches should pull their goalies much sooner. Suppose that if both teams are at full strength, each team scores an average of .05 goals per minute. Also suppose that if you pull your goalie you score an average of .08 goals per minute while your opponent scores an average of .12 goals per minute. Suppose you are one goal behind with five minutes left in the game. Consider the following two strategies:

Strategy 1: Pull your goalie if you are behind at any point in the last five minutes of the game; put him back in if you tie the score or go ahead.

Strategy 2: Pull your goalie if you are behind at any point in the last minute of the game; put him back in if you tie the score or go ahead.

Which strategy maximizes your chance of winning or tying the game? Simulate the game using ten-second increments of time. Use the `@RISKBINOMIAL` function to determine whether a team scores a goal in a given ten-second segment. This is OK because the probability of scoring two or more goals in a ten-second period is near 0.

References

Kelly, J. 1956. "A New Interpretation of Information Rate." *Bell System Technical Journal* 35: 917–926.

Marcus, A. 1990. "The Magellan Fund and Market Efficiency." *Journal of Portfolio Management* (Fall): 85–88.

Morrison, D., and R. Wheat. 1984. "Pulling the Goalie Revisited." *Interfaces* 16 (no. 6): 28–34.

CHAPTER 14

Using Simulation to Determine Optimal Maintenance Policies

In this section, we show how to use simulation to determine optimal maintenance policies. Our example also introduces the reader to the useful spreadsheet LOOKUP table command. The following example illustrates the idea.

Example 14.1

At the beginning of each week a machine is in one of four conditions: 1 = excellent, 2 = good, 3 = average, and 4 = bad. The weekly revenue earned by a machine is as follows: excellent machine, $100; good machine, $80; average machine, $50; bad machine, $10. After observing the condition of the machine at the beginning of the week, we have the option (for the cost of $200) of instantaneously replacing the machine with an excellent machine. The quality of the machine deteriorates over time, as shown in Table 14.1.

Table 14.1 Description of Machine Deterioration

Present Machine State	Probability That Machine Begins Next Week As			
	Excellent	*Good*	*Average*	*Bad*
Excellent	.7	.3		
Good		.7	.3	
Average			.6	.4
Bad				1.0 (until replaced)

Four maintenance policies are under consideration:

Policy 1: Never replace a machine.

Policy 2: Immediately replace a bad machine.

Policy 3: Immediately replace a bad or average machine.

Policy 4: Immediately replace a bad, average, or good machine.

Simulate each of these policies for 50 weeks in an attempt to determine the policy that maximizes expected weekly profit. Assume we begin Week 1 with an excellent machine.

Solution Note that Policy 1 will have the smallest weekly expected replacement costs, but it will also have the smallest weekly expected revenue. Also, Policy 4 will have the largest weekly expected replacement costs and the largest weekly expected revenue. Thus it is not obvious which policy is best.

Simulating Policy 1

We begin by simulating Policy 1 (see Figure 14.1 and file Mainnr.wk4 or Mainnr.xls). We proceed as follows:

Step 1 In the cell range B3:B6 we enter the weekly profit corresponding to the type of machine present at the beginning of the week. For example, if we begin the week with a Type 2 (good) machine, our weekly profit is $80. In rows 9–58, we simulate 50 weeks of machine replacement.

Step 2 In B9 we enter a 1 because we begin Week 1 with an excellent machine.

Step 3 In cell C9 we use the formula

```
Excel: =VLOOKUP(B9,$A$3:$B$6,2)

Lotus: @VLOOKUP(B9,$A$3..$B$6,1)
```

to compute the Week 1 profit. This formula says to find the week's profit by looking in the second column of the "lookup table range" (A3:B6) under the row in the lookup range corresponding to the entry in B9. Copying this formula to the range C9:C58 generates the profit for each week.

	A	B	C	D	E	F	G	H
1	NEVER	REPLACE	POLICY	MEAN		Figure 14.1		
2	TYPE	RB PROFIT		PROFIT				
3	1	100		22.2				
4	2	80						
5	3	50		Simulation Results for				
6	4	10						
7				Iterations= 100				
8	WEEK	TYPE	PROFIT	Simulations= 1				
9	1	1	100	# Input Variables= 147				
10	2	1	100	# Output Variables= 2				
11	3	1	100	Sampling Type= Latin Hypercube				
12	4	2	80	Runtime= 00:00:25				
13	5	2	80					
14	6	3	50	Summary Statistics				
15	7	3	50					
16	8	3	50	Cell	Name	Minimum	Mean	Maximum
17	9	3	50					
18	10	3	50					
19	11	4	10	!D3	/PROFIT	14	23.468	46
51	43	4	10					
52	44	4	10					
53	45	4	10					
54	46	4	10					
55	47	4	10					
56	48	4	10					
57	49	4	10					
58	50	4	10					

Figure 14.1 Simulation for Never Replace Machine Rule

Step 4 To generate the type of machine present for each week we enter the following formula into cell B10:

```
Excel: =IF((B9=1),RISKDISCRETE({1,2},{.7,.3}),
IF((B9=2), RISKDISCRETE({2,3},{.7,.3}),
IF((B9=3),RISKDISCRETE({3,4},{.6,.4}),4)))
```

```
Lotus: @IF((B9=1),@RISKDISCRETE(1,,2,.7,.3),
@IF((B9=2), @RISKDISCRETE(2,3,.7,.3),
@IF((B9=3)), @RISKDISCRETE(3,4,.6,.4),4)))
```

This formula ensures, for example, that if we begin the week with a Type 2 machine, then there is a .7 chance that we will begin the next week with a Type 2 machine and a .3 chance that we will begin the next week with a Type 3 machine. Copying this formula to the range B10:B58 generates the type of machine on hand at the beginning of each of the first 50 weeks.

Step 5 In cell D3 we compute the average profit for the 50 week period with the formula

```
Excel: =AVERAGE(C9:C58)

Lotus: @AVG(C9..C58)
```

Step 6 We now let cell D3 be our output range and use @RISK to run 100 iterations of the spreadsheet. We find that Policy 1 has an average profit of $23.47 per week.

Simulating Policy 2

We now set up the spreadsheet (Figure 14.2 and file Maintrb.wk4 or Maintrb.xls) that is used to estimate the average weekly profit for Policy 2 (we leave Policy 3 and Policy 4 as problems). We proceed as follows:

Step 1 In cell B6 we enter –100. This is because if we begin a week with a Type 4 (bad) machine we will replace it and our weekly profit will be $100 – $200 = –$100. Our formulas in column C remain the same. (Why?)

Step 2 To generate the type of machine present at the beginning of each week we enter in cell B9 the formula

```
Excel: =IF((B9=1),RISKDISCRETE({1,2},{.7,.3}),
IF((B9=2),RISKDISCRETE({2,3},{.7,.3}),
IF((B9=3),RISKDISCRETE({3,4},{.6,.4}),
RISKDISCRETE({1,2},{.7,.3}))))

Lotus: @IF((B9=1),@RISKDISCRETE(1,2,.7,.3),
@IF((B9=2), @RISKDISCRETE(2,3,.7,.3),
@IF((B9=3)), @RISKDISCRETE(3,4,.6,.4),
@RISKDISCRETE(1,2,.7,.3))))
```

Note that this formula ensures that if the week begins with a Type 4 machine, there is a .7 chance that the next week will begin with a Type 1 machine and a .3 chance that the next week will begin with a Type 2 machine. This is because Policy 2 replaces a bad machine with an excellent machine. Copying this formula to the range B9:B58 generates the machine type present at the beginning of each week.

	A	B	C	D	E	F	G	H
1	**Replace only Bad Machine**			MEAN	**Figure 14.2**			
2	TYPE	RB PROFIT		PROFIT				
3	1	100		69				
4	2	80						
5	3	50		Simulation Results for				
6	4	-100						
7				Iterations= 100				
8	WEEK	TYPE	PROFIT	Simulations= 1				
9	1	1	100	# Input Variables= 196				
10	2	1	100	# Output Variables= 1				
11	3	1	100	Sampling Type= Latin Hypercube				
12	4	1	100	Runtime= 00:00:49				
13	5	2	80					
14	6	3	50	Summary Statistics				
15	7	4	-100					
16	8	1	100	Cell	Name	Minimum	Mean	Maximum
17	9	1	100					
18	10	2	80	D3	/PROFIT	47.8	59.44	74.2
19	11	3	50					
51	43	1	100					
52	44	1	100					
53	45	1	100					
54	46	1	100					
55	47	1	100					
56	48	2	80					
57	49	2	80					
58	50	2	80					

Figure 14.2 Simulation for Replace only Bad Machine Rule

Step 3 Running a simulation with cell D3 as the output cell yields an estimate of average weekly profit to be $59.44.

Remarks

1. Note that Excel cannot handle more than seven nested `IF`s in the same cell. Also, if an `IF` statement in Excel contains more than 200 characters, @RISK may have trouble running a simulation.
2. In the Excel version of @RISK you cannot use `VLOOKUPS` within a `=RISKDISCRETE` function.

What Is the Best Policy?

Simulating Policies 3 and 4 (see Problems 14.1 and 14.2) we find that the average weekly profit for Policy 3 is $60.91 and the average weekly profit for Policy 4 is $41.20. In conclusion, Policy 3 appears to maximize expected weekly profit, with Policy 2 close behind. Policies 1 and 4 are vastly inferior.

Problems

Group A

14.1 Simulate Policy 3 of Example 14.1.

14.2 Simulate Policy 4 of Example 14.1.

Group B

14.3 An important machine is known to never last more than four months. During its first month of operation it fails 10% of the time. If the machine completes its first month of operation, then it fails during its second month of operation 20% of the time. If the machine completes its second month of operation, then it will fail during its third month of operation 50% of the time. If the machine completes its third month of operation, then it is sure to fail by the end of the fourth month. At the beginning of each month we must decide whether to replace our machine with a new machine. It costs $500 to purchase a new machine, but if a machine fails during a month, we incur a cost of $1000 (as the result of factory downtime) and must replace the machine (at the beginning of the next month) with a new machine. Three maintenance policies are under consideration:

Policy 1: Plan to replace a machine at the beginning of its fourth month of operation.

Policy 2: Plan to replace a machine at the beginning of its third month of operation.

Policy 3: Plan to replace a machine at the beginning of its second month of operation.

Which policy will give the lowest average monthly cost?

As an example of what might happen, suppose we make a planned replacement of a machine at the beginning of a month. If this machine does not fail during the month, we incur $500 in costs. If the machine does fail during the month we incur $500 + $1000 + $500 = $2000 in costs for the current month.

CHAPTER 15

Using the Weibull Distribution to Model Machine Life

Many companies have to decide whether to replace a machine before it fails. The reason for replacing a machine is that a planned replacement of a machine may result in less downtime than an unplanned replacement. This may more than compensate for the fact that a planned replacement strategy will result in higher replacement costs than a strategy of only replacing a machine when it fails.

In order to use @RISK to analyze machine replacement decisions, we need to have a distribution to model the lifetime of a machine. The **Weibull distribution** is the most widely used distribution for modeling the life of a machine. The Weibull distribution is specified by two parameters, α and β. If $\beta < 1$, the machine's lifetime has the decreasing failure rate property. This means that the longer the machine has operated, the less likely it is to fail during the next few seconds. If $\beta > 1$, the machine's lifetime has the increasing failure rate property. This means that the longer the machine has operated, the more likely it is to fail during the next few seconds. If $\beta = 1$, then the machine's lifetime is exponential and has the constant failure rate property or lack of memory. The lack of memory property implies that no matter how long a machine has been operating, its chance of failing during the next few seconds remains the same.

To generate a machine lifetime with @RISK from a Weibull distribution with given parameters α and β enter the statement

```
Excel: =RISKWEIBULL(α,β)

Lotus: @RISKWEIBULL(α,β)
```

	A	B	C	D	E	F	G
1		Estimating Weibull		Figure 15.1			
2		Distribution Parameters					
3							
4		Mean time to failure		46000			
5		Variance of time to Failure		1.56E+08			
6		Second Moment of failure time		2.27E+09			
7		Second moment/(mean)^2		1.073842		Beta	50608.53
8		Alpha				Alpha	4.2

Figure 15.1 Estimating Parameters for Weibull Distribution

Given the mean and variance of a machine's lifetime, the file Weibest.xls (see Figure 15.1) enables you to estimate α and β for the machine's lifetime. Simply enter the mean lifetime of the machine in cell D4 and the variance of the machine's lifetime in cell D5. Then the values of α and β can be read in cells G7 and G8. For example, if the mean lifetime of a machine is 46,000 hours and the variance of its lifetime is 156,000,000 hours, then we estimate that $\alpha = 4.2$ and $\beta = 50{,}608.5$.

The following example shows how to use the Weibull distribution to analyze equipment replacement decisions.

Example 15.1

The lifetime of a stamping press used to manufacture truck bodies follows a Weibull distribution with $\alpha = 6$ and $\beta = 60$. Every hour the drill press is down costs the plant \$200. It costs \$5000 to purchase a new drill press. If we make a planned replacement of a drill press, the press will be down for 2 hours; if we need to replace a failed drill press, the press will be down for 20 hours. We are considering introducing a strategy of letting a drill press run for a specified number of hours, and then replacing it. What strategy will minimize our expected cost per hour?

Solution Our spreadsheet is in Figure 15.2 (see file Weibull.wk4 or Weibull.xls).

To model a given replacement strategy we will simulate the operation of the drill press for a substantial length of time (say 1000 hours) and determine a planned replacement strategy that minimizes estimated expected costs (downtime and replacement) of operating a press for 1000 hours. We proceed as follows:

Step 1 Begin at Time 0 and determine (in columns B and C) the lifetime of the machine and the next planned replacement.

Step 2 Determine (in column D) whether the 1000-hour time limit has been exceeded. If it has, no more costs should be incurred.

Step 3 Determine the number of downtime hours associated with the replacement of the current press (column E).

	A	B	C	D	E	F	G	H	I	J	K	L
1	Determining optimal		Figure 15.2									
2	replacement policy											
3												
4	Cost per hour of down time		$ 200.00									
5	Cost of replacement		$ 5,000.00									
6	Hours for planned replacement		2									
7	Hours for unplanned replacement		20									
8	Weibull alpha		6									
9	Weibull beta		60					Total Cost				
10	Planned replacement interval		20	Totals		230000	18400	248400				
11					Down-Time	Replacement	Down-time					
12	Time	Next Failure	Next planned replacement	Over?	Hours	Cost	Cost					
13	0	48.03981944	20	1	2	5000	400					
14	22	76.20961878	42	1	2	5000	400					
15	44	104.1744552	64	1	2	5000	400	Iterations= 50				
16	66	128.0860153	86	1	2	5000	400	Simulations= 5				
17	88	124.2092545	108	1	2	5000	400	# Input Variables= 51				
18	110	157.9870459	130	1	2	5000	400	# Output Variables= 1				
19	132	177.9225834	152	1	2	5000	400	Sampling Type= Latin Hypercube				
20	154	209.9201987	174	1	2	5000	400	Runtime= 00:01:14				
21	176	245.5401283	196	1	2	5000	400					
22	198	256.3479085	218	1	2	5000	400	Summary Statistics				
23	220	273.3901726	240	1	2	5000	400					
24	242	303.9503448	262	1	2	5000	400	Cell	Name	Minimum	Mean	Maximum
25	264	310.3859652	284	1	2	5000	400					
26	286	356.355342	306	1	2	5000	400	H10	(Sim#1) Totals/Total	246600	248292	248400
27	308	373.6642643	328	1	2	5000	400	H10	(Sim#2) Totals/Total	129600	135648	145800
28	330	390.1576566	350	1	2	5000	400	H10	(Sim#3) Totals/Total	100800	119016	133200
29	352	393.9915924	372	1	2	5000	400	H10	(Sim#4) Totals/Total	113400	123300	135000
30	374	403.9445382	394	1	2	5000	400	H10	(Sim#5) Totals/Total	108000	123300	135000
41	616	689.8810841	636	1	2	5000	400					
42	638	690.0299949	658	1	2	5000	400					
43	660	709.4283516	680	1	2	5000	400					
44	682	723.9128883	702	1	2	5000	400					
45	704	763.5621828	724	1	2	5000	400					
46	726	784.2952103	746	1	2	5000	400					
47	748	806.8835301	768	1	2	5000	400					
48	770	819.1087203	790	1	2	5000	400					
49	792	854.0614114	812	1	2	5000	400					
50	814	868.0639787	834	1	2	5000	400					
51	836	889.2785493	856	1	2	5000	400					
52	858	908.5532082	878	1	2	5000	400					
53	880	933.5074928	900	1	2	5000	400					
54	902	962.8994034	922	1	2	5000	400					
55	924	972.2140148	944	1	2	5000	400					
56	946	1012.911889	966	1	2	5000	400					
57	968	994.7246505	988	1	2	5000	400					
58	990	1047.56275	1010	1	2	5000	400					
59	1012	1078.505036	1032	0	2	0	0					

Figure 15.2 Optimal Machine Replacement with the Weibull Distribution

Step 4 Determine the replacement and downtime cost associated with the replacement of the current press (columns F and G).

Step 5 Determine the time the next press will begin operation (column A) by the relation

(Time operation of next press begins) =
Minimum (Time Current Press Fails, Time of Next Planned Replacement)
+ (Downtime Associated with Current Machine) **(15.1)**

If we type the formula

```
Excel: =RISKWEIBULL(6,60)

Lotus: @RISKWEIBULL(6,60)
```

anywhere in our spreadsheet, we find (if the Monte Carlo option is off) that the mean life of a machine is 55.6 hours. With this in mind, it seems reasonable to begin by experimenting with replacing the press after 20, 40, 60, 80, or 100 hours of operation. We now proceed as follows:

Step 1 In the cell range C4:C10 we enter the inputs to our problem. The possible planned replacement intervals of 20, 40, 60, 80, and 100 hours are implemented in cell C10 with the statement

```
Excel: =RISKSIMTABLE({20,40,60,80,100})

Lotus: @RISKSIMTABLE(20,40,60,80,100)
```

Step 2 Enter a 0 in cell A13 to indicate the start of the simulation.

Step 3 In cell B13 we compute the time the first drill press fails by generating the lifetime of the first press with a Weibull distribution and adding this lifetime to the time the press begins operation (0). Thus we enter in cell B13 the formula

```
Excel: =A13+RISKWEIBULL(C$8,C$9)

Lotus: +A13+@RISKWEIBULL(C$8,C$9)
```

Step 4 We now compute the instant of the next planned replacement by the relationship

(Next Planned Replacement) =
(Time of Current Machine's Start of Operation)
+ (Planned Replacement Interval)

To do this enter in cell C13 the formula

```
Excel: =A13+C$10

Lotus: +A13+C$10
```

Step 5 In cell D13 we determine whether 1000 hours has been exceeded (indicated by a 0) with the formula

```
Excel: =IF(A13<1000,1,0)
Lotus: @IF(A13<1000,1,0)
```

Step 6 In cell E13 we determine the number of downtime hours associated with the replacement of the first press with the formula

```
Excel: =IF(B13<C13,C$7,C$6)
Lotus: @IF(B13<C13,C$7,C$6)
```

This records a downtime of 20 hours if the machine is replaced due to failure and a downtime of 2 hours if the machine is replaced due to a planned replacement.

Step 7 In cell F13 we compute the replacement cost associated with the first press with the formula

```
Excel: =D13*C$5
Lotus: +D13*C$5
```

This formula will pick up the replacement cost only if the simulation has not yet run 1000 hours.

Step 8 In cell G13 we compute the downtime cost associated with the first press via the formula

```
Excel: =D13*E13*C$4
Lotus: +D13*E13*C$4
```

This formula will pick up the downtime cost for the press only if the simulation has not yet run 1000 hours.

Step 9 In cell A14 we determine the time the second press begins operation by operationalizing (15.1) with the formula

```
Excel: =MIN(B13,C13)+E13
Lotus: @MIN(B13,C13)+E13
```

Step 10 Our minimum planned replacement interval is 20 hours. Accounting for the 2 hours of downtime associated with each planned replacement, 1000 hours of operation would require at most 1000/22 = 46 press replacements. Therefore we need only copy our spreadsheet down to row 59. We now copy the formula in A14 to the range A14:A59. Next we copy the formulas in the range B13:G13 to the range B14:G59.

Step 11 In cell F10 we compute the total replacement cost incurred in 1000 hours with the formula

```
Excel: =SUM(F13:F59)
Lotus: @SUM(F13:F59)
```

Copying this formula to cell G10 computes the total downtime cost.

Step 12 Finally, in cell H10 we compute the *total* cost incurred during 1000 hours of operation with the formula

```
Excel:  =SUM(F10:G10)
Lotus:  @SUM(F10..G10)
```

Next we ran 50 iterations of five simulations to determine which planned replacement interval minimizes total cost for 1000 hours. We find that Simulation 3 (corresponding to replacing a press after 60 hours of operation) minimizes expected cost.

Problems

Group A

15.1 Suppose we decide to replace the drill press in Example 15.1 every 100 hours. That is, at time 100, 200, 300, etc., we will replace the drill press, no matter how long it has been in operation. Estimate the expected cost per hour incurred by this replacement policy.

Hint: You might use the Excel `CEILING` or Lotus `ROUNDUP` function to help compute the time of the next planned replacement.

15.2 A space capsule is scheduled to go on a 100-week flight. It sends back valuable signals using a battery-powered transmitter. The lifetime of a battery follows a Weibull distribution with $\alpha = 2$ and $\beta = 60$. How many batteries should we take on the flight in order to be 99% sure that there is a battery working during the entire flight? Assume the capsule is set up so that once the first battery fails, the second battery automatically kicks in, etc.

15.3 Consider a drill press containing three drill bits. The current policy (called **individual replacement**) is to replace a drill bit when it fails. The firm is considering changing to a **block replacement policy** in which all three drill bits are replaced whenever a single drill bit fails. Each time the drill press is shut down the cost is \$100. A drill bit costs \$50, and the variable cost of replacing a drill bit is \$10. Assume that the time to replace a drill bit is negligible. Assume that the time until failure for a drill bit follows an exponential distribution with a mean of 100 hours. This can be modeled in @RISK with the formula

```
Excel:  =RISKEXPON(100)
Lotus:  @RISKEXPON(100)
```

Determine which replacement policy (block or individual replacement) should be implemented.

15.4 Redo Problem 15.3 under the assumption that the time to failure of a drill bit follows a Weibull distribution with $\alpha = 6$ and $\beta = 60$.

CHAPTER 16

Simulating Stock Prices and Options

In this chapter we show how you can use @RISK to simulate the behavior of stock prices. We also show how you can use @RISK to analyze the value of portfolios consisting of derivative securities (such as call and put options) whose value depends on the value of an underlying stock(s).

16.1 Modeling the Price of a Stock

Most financial models of stock prices assume that the stock's price follows a **log-normal distribution**. Essentially this means that the logarithm of the stock's price at any time is a normally distributed random variable. Let

P_0 = Price of stock now (at Time 0); this is known.

t = Any future time (measured in years).

P_t = Price of stock at Time t; it is a random variable and its value is not known until Time t!

Z = A standard normal random variable (having mean 0 and standard deviation 1).

μ = Mean percentage growth rate of stock (per year) expressed as a decimal.

σ = Standard deviation of the growth rate of stock (per year) expressed as a decimal.

We will soon show how to estimate μ and σ from historical data. Assuming that we have good estimates of μ and σ most finance practitioners assume that P_t may be modeled by

$$P_t = P_0 * \exp[(\mu - .5 * \sigma^2) * t + \sigma * Z * t^{.5}] \quad \textbf{(16.1)}$$

Note that ln P_t is indeed normally distributed, so P_t does follow a lognormal distribution. It is easy to use (16.1) to model the evolution of a stock's price with @RISK. We begin by showing how to estimate μ and σ from historical data.

16.2 Estimating the Mean and Standard Deviation of Stock Returns from Historical Data

Suppose the current price of a stock is $25.00 and the monthly closing prices of a stock during the next 12 months are as follows:

Table 16.1 Monthly Closing Prices for a Stock

Month	Closing Price	Month	Closing Price
1	$24.70	7	$23.94
2	$23.70	8	$24.37
3	$22.90	9	$24.99
4	$22.81	10	$26.09
5	$22.89	11	$26.14
6	$22.56	12	$26.90

The spreadsheet in Figure 16.1 (file Hisvar.wk4 or Hisvar.xls) shows how to use this data to estimate the mean and standard deviation of the stock's monthly return. Essentially, you compute ln(1 + Percentage Return) for each month, and take the mean and standard deviation of these twelve numbers. From Figure 16.1 we find that $\mu_{month} = .0061$ and $\sigma_{month} = .0288$. To convert monthly means and standard deviations to annual values, proceed as follows:

$$\mu_{annual} = 12 * \mu_{monthly} = 12 * (.0061) = .073 \quad \textbf{(16.2)}$$

$$\sigma_{annual} = 12^{.5} * \sigma_{monthly} = (3.464) * (.0288) = .0998 \quad \textbf{(16.3)}$$

Thus the stock has a mean average annual return of 7.3% and a standard deviation of 9.98%.

	A	B	C	D	E
1	FIGURE	16.1			
2					
3	MONTH	CLOSING	RETURN	1+RETURN	LN(1+RETURN)
4	0	$25.00			
5	1	$24.70	-0.0120	0.9880	-0.012072581
6	2	$23.70	-0.0405	0.9595	-0.041328195
7	3	$22.90	-0.0338	0.9662	-0.034338138
8	4	$22.81	-0.0039	0.9961	-0.003937874
9	5	$22.89	0.0035	1.0035	0.003501098
10	6	$22.56	-0.0144	0.9856	-0.014521707
11	7	$23.94	0.0612	1.0612	0.059372273
12	8	$24.37	0.0180	1.0180	0.017802168
13	9	$24.99	0.0254	1.0254	0.025122877
14	10	$26.09	0.0440	1.0440	0.043076354
15	11	$26.14	0.0019	1.0019	0.001914609
16	12	$26.90	0.0291	1.03	0.028659578
17				MEAN=	0.006104205
18				STDEV=	0.028835938

Figure 16.1 Estimating Mean and Standard Deviation of a Stock from Historical Data

Remarks

1. Formula (16.2) follows from the fact that the expected value of a sum of random variables equals the sum of the expected values of the random variables.

2. Formula (16.3) follows from the fact that the variance of a sum of independent random variables equals the sum of the variances of the individual random variables.

16.3 What Is an Option?

A **European option** on a stock gives the owner of the option the right to buy (if the option is a **call** option) or sell (if the option is a **put** option) for a particular price one share of a stock on a particular date. The price at which an option holder can buy or sell a stock is called the **exercise price**. The date by which the option must be used (or "exercised") is called the **expiration date.**

For example, suppose that a stock is currently selling for $50 and you purchase a call option with an exercise price of $56 and a three-month expiration date. What cash flows do you obtain from this option? Let S_T = price of stock at the expiration date. Then your cash flows from this option (ignoring cost of buying option) are as follows:

Table 16.2 Cash Flow from Call Option

Case	Cash Flow
$S_T \leq 56$ (out of the money)	0
$S_T > 56$ (in the money)	$S_T - 56$

This relation follows because if three months from now the price of the stock exceeds $56, we may buy one share of the stock for $56 and sell it on the open market for its current price, thereby earning a profit of $S_T - 56$. If the price of the stock at the expiration date is less than $56, then we cannot make any money by purchasing the option. In short

$$\text{Cash Flow from Call Option} = \text{Maximum}\{0, S_T - 56\} \qquad \textbf{(16.4)}$$

For a European put option the reader should be able to show that

Table 16.3 Cash Flow from Call Option

Case	Cash Flow
$S_T \leq 56$ (in the money)	$56 - S_T$
$S_T > 56$ (out of the money)	0

Thus

$$\text{Cash Flow from Put Option} = \text{Max}(56 - ST, 0) \qquad \textbf{(16.5)}$$

16.4 Pricing a Call Option

Since option trading is a multibillion-dollar business, it is natural to ask what is a fair price for a given option. Black and Scholes (1973) were the first to come up with a formula to price options. Their work revolutionized investments and corporate finance. In 1979 Cox, Ross, and Rubenstein came up with a different method for pricing options (the arbitrage pricing model). Their work showed that the price of an option must be the expected *discounted* (continuously at the risk-free rate) *value of the cash flows from an option on a stock having the same variance as the stock on which the option is written and growing at the risk-free rate of interest.* If the option is not priced according to this rule, traders can make arbitrage profits. Fortunately, both the Black-Scholes and Cox-Ross-Rubenstein approaches yield the same option price! Note that the price of the option does not depend on the growth rate of the stock!

Let's use @RISK to price a European call option.

Example 16.1a

A share of Eli Daisy currently sells for \$42. A European call option with an expiration date of six months and an exercise price of \$40 is available. The stock has an annual standard deviation of 20%. The stock price has tended to increase at a rate of 15% per year. The risk-free rate is 10% per year. What is a fair price for this option?

Solution The @RISK simulation for this situation is in Figure 16.2 (file Option.wk4 or Option.xls). We begin by simulating the price of the stock in six months. We use (16.1) to do this. We proceed as follows:

Step 1 Enter the relevant information in the cell range D3:D8 (see Figure 16.2a).

Step 2 To generate the stock price six months from now we enter in cell D10

```
Excel:  =D3*EXP((D7-.5*D6^2)*D8+D6
*RISKNORMAL(0,1)*D8^.5)

Lotus:  +D3*@EXP((D7-.5*D6^2)*D8+D6
*RISKNORMAL(0,1)*D8^.5)
```

Note that we assume that the stock grows at the risk-free rate, *not the stock's growth rate*. The Black-Scholes derivation of the call option price shows that the stock's growth rate does not influence the price of a call option. (See Problem 16.8 for an intuitive explanation of why a stock's growth rate does not affect the value of an option on the stock.)

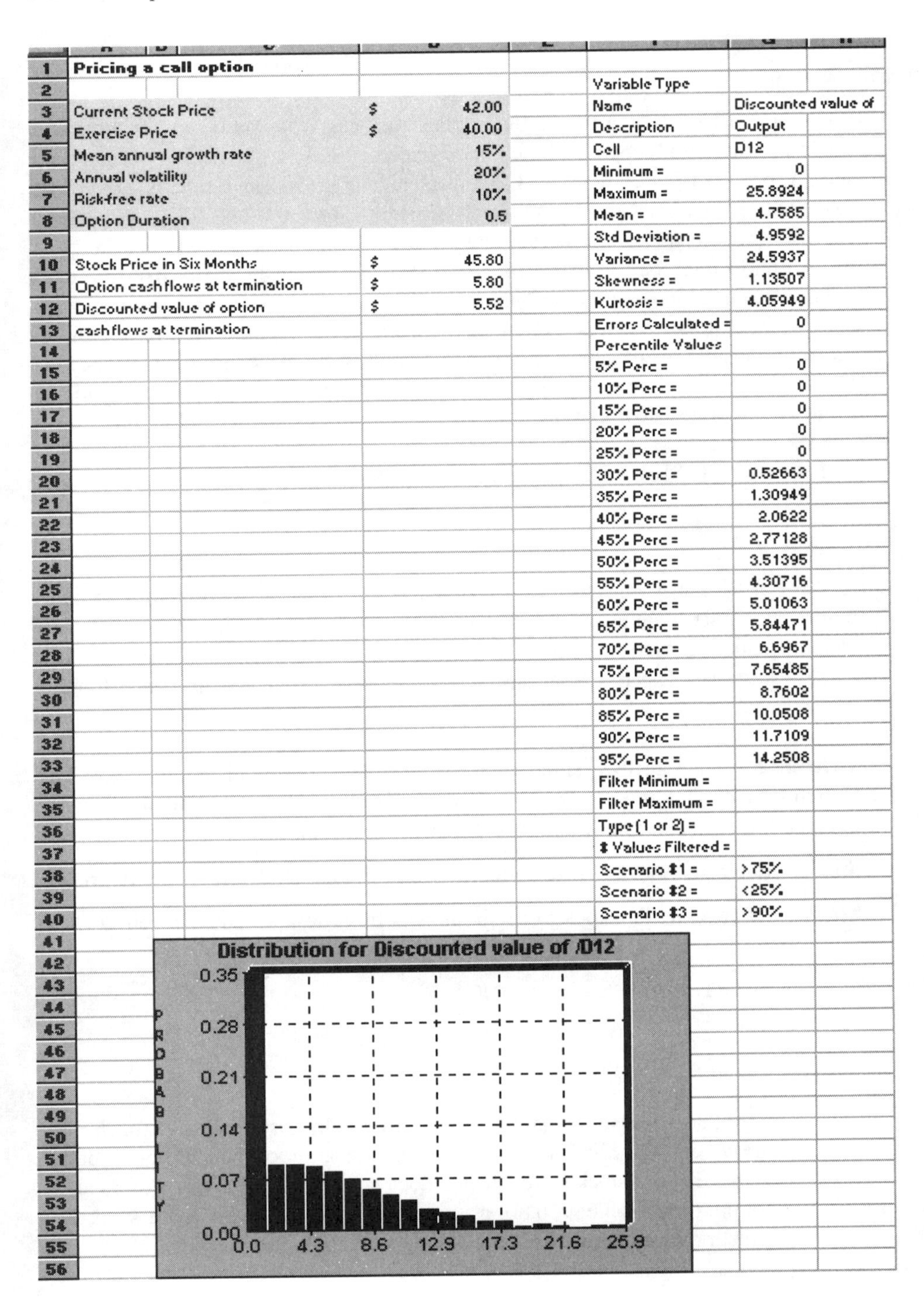

Row	Item		Value
1	Pricing a call option		
2			
3	Current Stock Price	$	42.00
4	Exercise Price	$	40.00
5	Mean annual growth rate		15%
6	Annual volatility		20%
7	Risk-free rate		10%
8	Option Duration		0.5
9			
10	Stock Price in Six Months	$	45.80
11	Option cash flows at termination	$	5.80
12	Discounted value of option	$	5.52
13	cash flows at termination		

Row	F	G
2	Variable Type	
3	Name	Discounted value of
4	Description	Output
5	Cell	D12
6	Minimum =	0
7	Maximum =	25.8924
8	Mean =	4.7585
9	Std Deviation =	4.9592
10	Variance =	24.5937
11	Skewness =	1.13507
12	Kurtosis =	4.05949
13	Errors Calculated =	0
14	Percentile Values	
15	5% Perc =	0
16	10% Perc =	0
17	15% Perc =	0
18	20% Perc =	0
19	25% Perc =	0
20	30% Perc =	0.52663
21	35% Perc =	1.30949
22	40% Perc =	2.0622
23	45% Perc =	2.77128
24	50% Perc =	3.51395
25	55% Perc =	4.30716
26	60% Perc =	5.01063
27	65% Perc =	5.84471
28	70% Perc =	6.6967
29	75% Perc =	7.65485
30	80% Perc =	8.7602
31	85% Perc =	10.0508
32	90% Perc =	11.7109
33	95% Perc =	14.2508
34	Filter Minimum =	
35	Filter Maximum =	
36	Type (1 or 2) =	
37	# Values Filtered =	
38	Scenario #1 =	>75%
39	Scenario #2 =	<25%
40	Scenario #3 =	>90%

Figure 16.2 Pricing a Call Option

Figure 16.2a

	A	B	C	D
1	**Pricing a call option**			
2				
3	Current Stock Price			$ 42.00
4	Exercise Price			$ 40.00
5	Mean annual growth rate			15%
6	Annual volatility			20%
7	Risk-free rate			10%
8	Option Duration			0.5
9				
10	Stock Price in Six Months			$ 45.80
11	Option cash flows at termination			$ 5.80
12	Discounted value of option			$ 5.52
13	cash flows at termination			

Variable Type		
Name	Discounted value of	
Description	Output	
Cell	D12	
Minimum =	0	
Maximum =	25.8924	
Mean =	4.758498	
Std Deviation =	4.9592	
Variance =	24.59367	
Skewness =	1.135074	
Kurtosis =	4.059486	
Errors Calculated =	0	
Percentile Values		
5% Perc =	0	
10% Perc =	0	
15% Perc =	0	
20% Perc =	0	
25% Perc =	0	
30% Perc =	0.52663	
35% Perc =	1.309485	
40% Perc =	2.062199	
45% Perc =	2.771283	
50% Perc =	3.513945	
55% Perc =	4.307158	
60% Perc =	5.01063	
65% Perc =	5.844711	
70% Perc =	6.696697	
75% Perc =	7.65485	
80% Perc =	8.760197	
85% Perc =	10.0508	
90% Perc =	11.7109	
95% Perc =	14.25078	
Filter Minimum =		
Filter Maximum =		
Type (1 or 2) =		
# Values Filtered =		
Scenario #1 =	>75%	
Scenario #2 =	<25%	
Scenario #3 =	>90%	

Figure 16.2b

Step 3 In cell D11 we compute the cash flows from the call option at expiration with the formula

```
Excel: =MAX(D10-D4,0)
Lotus: @MAX(D10-D4,0)
```

This formula embodies the essence of (16.4).

Step 4 In cell D12 we compute the discounted value of the cash flows at expiration. Note that if a cash flow is received at Time t and the discount rate (or risk-free rate) is r, then to discount the cash flows at Time t back to Time 0 they should be multiplied by e^{-rt}. Thus in D12 we enter the formula

```
Excel: =EXP(-D7*D8)*D11
Lotus: @EXP(-D7*D8)*D11
```

Step 5 We now choose cell D12 as our output cell and use @RISK to run 400 iterations of this spreadsheet (see Figure 16.2b). We found the average value of cell D12 to equal $4.758. According to the Black-Scholes option pricing formula, the price of this option should be $4.76! Thus we see that the simulation approach to option pricing has done the job!

Figure 16.2c

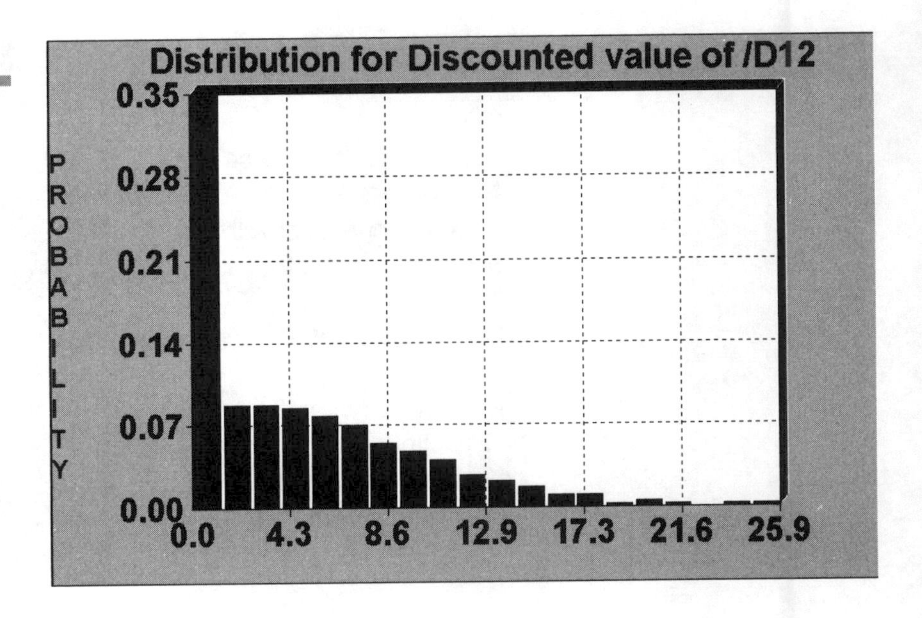

Remarks

1. Many options that are more complex (such as the so-called exotic and path-dependent options) than a call option can be valued with @RISK. See Chapter 17 for a discussion of path-dependent options. The key is to assume that the underlying stock grows at the risk-free rate. Then take the expected discounted value of the cash flows from the option. This will equal the "price" of the option found by the Cox-Ross-Rubenstein approach.

2. The histogram of the discounted cash flow from the call option given in Figure 16.2c shows that the discounted cash flow on the option exhibits high variability. For our simulation, the discounted cash flow from the option at termination ranged from $0 to almost $26. The large spike in the histogram in Figure 16.2c represents the iterations in which the option was out of the money (i.e., terminal stock price was less than exercise price of $40).

16.5 Analyzing a Portfolio of Investments

@RISK can easily be used to analyze the probability distribution of the return on a group of investments. Recall that the (percentage) return on a portfolio during a time period is given by

$$\text{Percentage Return on Portfolio} = \frac{(\text{Final Value of Portfolio}) - (\text{Initial Value of Portfolio})}{(\text{Initial Value of Portfolio})} \qquad \textbf{(16.6)}$$

Example 16.1b

A share of Eli Daisy currently sells for $42. We have just purchased a single share of this stock. A European call option with an expiration date of six months and an exercise price of $40 has also just been purchased. The stock has an annual standard deviation of 20%. The stock price has tended to increase at a rate of 15% per year. The risk-free rate is 10% per year. Use @RISK to analyze the percentage return on this portfolio over a six-month period.

Solution The key is to use formula (16.6) and remember that we should assume in our analysis that the stock grows at its actual growth rate, not the risk-free rate. Our work is presented in Figure 16.3 (Stpcall.xls or Stpcall.wk4). We proceed as follows (see Figure 16.3a):

Step 1 Since the stock grows at a rate of 15% per year we change the formula generating the stock price six months from now to

```
Excel: =D3*EXP((D5-.5*D6^2)*D8+D6
*RISKNORMAL(0,1)*D8^.5)

Lotus: +D3*@EXP((D5-.5*D6^2)*D8+D6
*RISKNORMAL(0,1)*D8^.5)
```

Step 2 In cell D15 we keep track of the ending value of our portfolio with the formula

```
Excel: =D10+D11

Lotus: +D10+D11
```

Of course, the ending value of the portfolio is random. Hit the F9 key a few times to see this!

Step 3 In cell D16 we compute the initial cost of the portfolio (the cost of a share of stock plus the cost of the call option, which we already know is 4.76) with the formula

```
Excel: =4.76+D3

Lotus: 4.76+D3
```

Step 4 In cell D17 we operationalize (16.6) and compute the percentage return on the portfolio with the formula

```
Excel: =(D15-D16)/D16

Lotus: (D15-D16)/D16
```

Step 5 We next selected cell D17 as our output cell and ran 400 iterations.

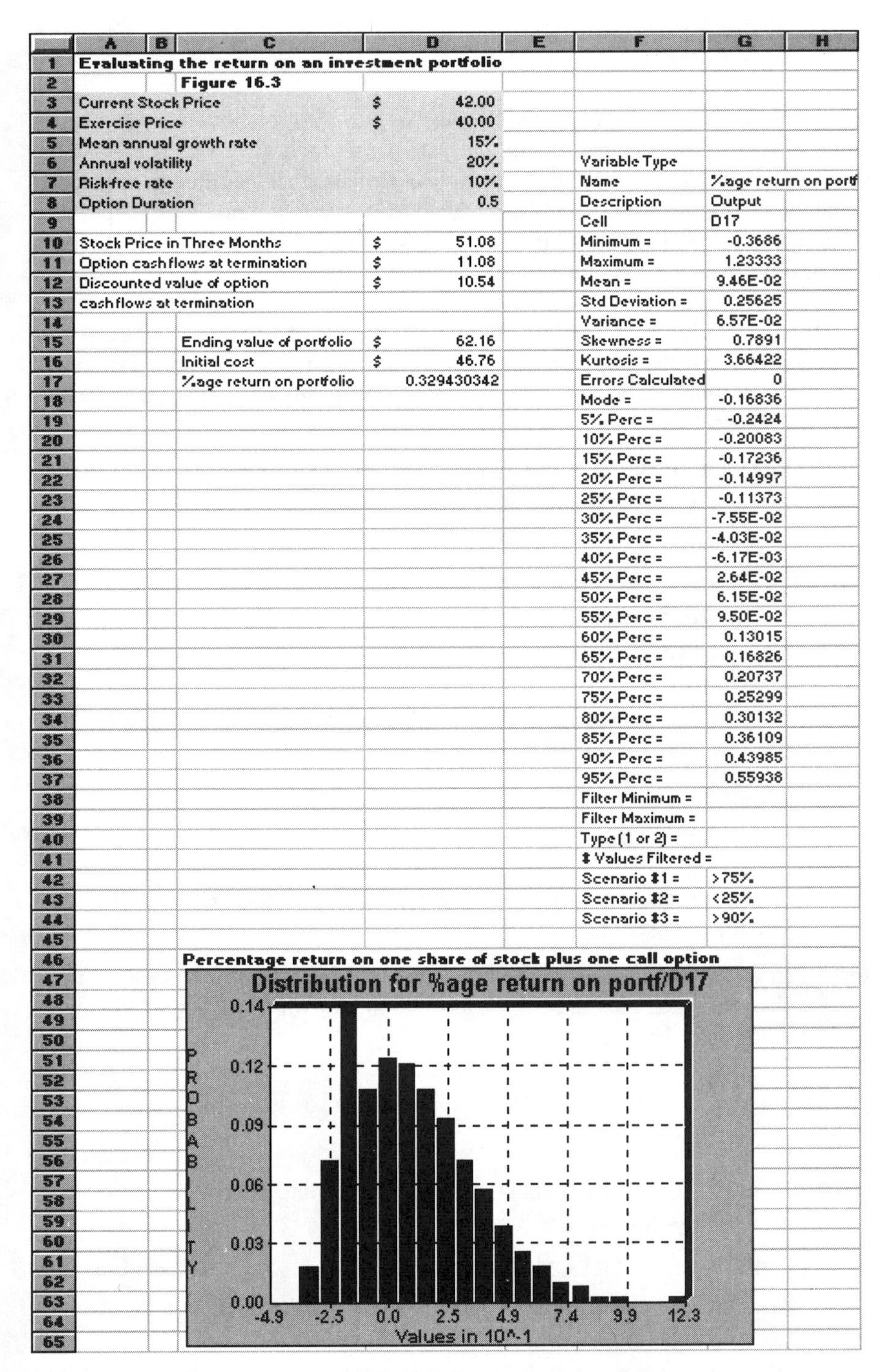

	A	B	C	D	E	F	G	H
1	**Evaluating the return on an investment portfolio**							
2			**Figure 16.3**					
3	Current Stock Price			$ 42.00				
4	Exercise Price			$ 40.00				
5	Mean annual growth rate			15%				
6	Annual volatility			20%		Variable Type		
7	Risk-free rate			10%		Name	%age return on portf	
8	Option Duration			0.5		Description	Output	
9						Cell	D17	
10	Stock Price in Three Months			$ 51.08		Minimum =	-0.3686	
11	Option cashflows at termination			$ 11.08		Maximum =	1.23333	
12	Discounted value of option			$ 10.54		Mean =	9.46E-02	
13	cashflows at termination					Std Deviation =	0.25625	
14						Variance =	6.57E-02	
15			Ending value of portfolio	$ 62.16		Skewness =	0.7891	
16			Initial cost	$ 46.76		Kurtosis =	3.66422	
17			%age return on portfolio	0.329430342		Errors Calculated	0	
18						Mode =	-0.16836	
19						5% Perc =	-0.2424	
20						10% Perc =	-0.20083	
21						15% Perc =	-0.17236	
22						20% Perc =	-0.14997	
23						25% Perc =	-0.11373	
24						30% Perc =	-7.55E-02	
25						35% Perc =	-4.03E-02	
26						40% Perc =	-6.17E-03	
27						45% Perc =	2.64E-02	
28						50% Perc =	6.15E-02	
29						55% Perc =	9.50E-02	
30						60% Perc =	0.13015	
31						65% Perc =	0.16826	
32						70% Perc =	0.20737	
33						75% Perc =	0.25299	
34						80% Perc =	0.30132	
35						85% Perc =	0.36109	
36						90% Perc =	0.43985	
37						95% Perc =	0.55938	
38						Filter Minimum =		
39						Filter Maximum =		
40						Type (1 or 2) =		
41						# Values Filtered =		
42						Scenario #1 =	>75%	
43						Scenario #2 =	<25%	
44						Scenario #3 =	>90%	
45								
46			**Percentage return on one share of stock plus one call option**					

Figure 16.3 Evaluating Percentage Return on a Portfolio

Figure 16.3a

	A	B	C	D
1	Evaluating the return on an investment portfolio			
2			Figure 16.3	
3	Current Stock Price			$ 42.00
4	Exercise Price			$ 40.00
5	Mean annual growth rate			15%
6	Annual volatility			20%
7	Risk-free rate			10%
8	Option Duration			0.5
9				
10	Stock Price in Six Months			$ 51.08
11	Option cash flows at termination			$ 11.08
12	Discounted value of option			$ 10.54
13	cash flows at termination			
14				
15			Ending value of portfolio	$ 62.16
16			Initial cost	$ 46.76
17			%age return on portfolio	0.329430342

Figure 16.3b

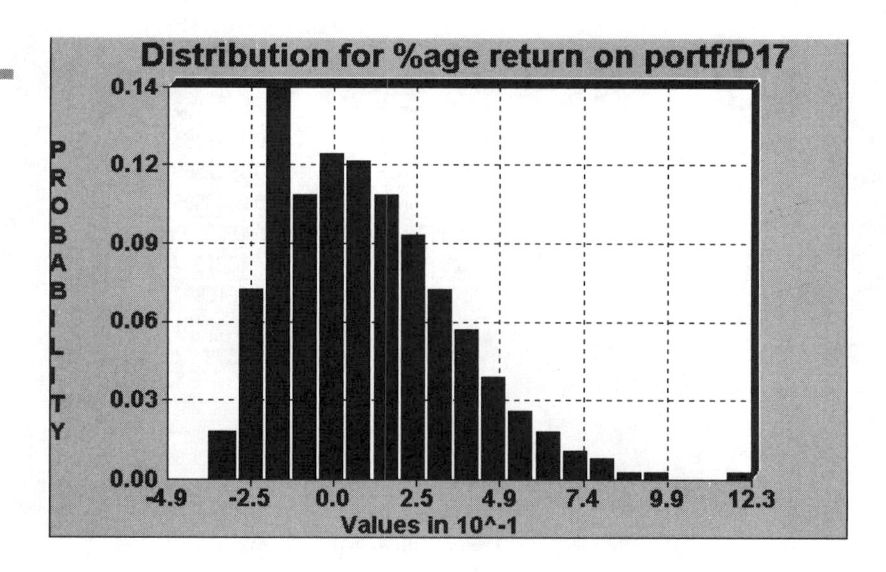

The results in Figures 16.3b and 16.3c indicate that we have a very risky portfolio. Our percentage return ranges from a loss of 37% to a gain of 123%. Our average return is 9.46% for the six-month period, but there is over a 40% chance that the portfolio will lose money! The reason the portfolio is so risky is that when the stock price goes up, both our share of stock and option do well; when the stock price goes down, both our

Figure 16.3c

Variable Type		
Name	%age return on portf	
Description	Output	
Cell	D17	
Minimum =	-0.3685974	
Maximum =	1.233329	
Mean =	9.46E-02	
Std Deviation =	0.2562477	
Variance =	6.57E-02	
Skewness =	0.789096	
Kurtosis =	3.664219	
Errors Calculated =	0	
Mode =	-0.1683566	
5% Perc =	-0.2424037	
10% Perc =	-0.2008321	
15% Perc =	-0.1723605	
20% Perc =	-0.149971	
25% Perc =	-0.1137341	
30% Perc =	-7.55E-02	
35% Perc =	-4.03E-02	
40% Perc =	-6.17E-03	
45% Perc =	2.64E-02	
50% Perc =	6.15E-02	
55% Perc =	9.50E-02	
60% Perc =	0.1301464	
65% Perc =	0.1682559	
70% Perc =	0.2073698	
75% Perc =	0.2529935	
80% Perc =	0.301317	
85% Perc =	0.3610862	
90% Perc =	0.4398451	
95% Perc =	0.5593764	
Filter Minimum =		
Filter Maximum =		
Type (1 or 2) =		
# Values Filtered =		
Scenario #1 =	>75%	
Scenario #2 =	<25%	
Scenario #3 =	>90%	

share of stock and option do poorly. Thus we have not *hedged* the risk associated with changes in stock price. A good way to hedge the risk associated with stock price changes is to buy a put option on a stock. See Problems 16.5 and 16.7.

Problems

Group A

16.1 A stock currently sells for \$100. The risk-free rate is 12% per year, and the stock's σ = 20%. By considering exercise prices of \$90, \$100, and \$110, show how the value of a call option depends on the option's exercise price. Assume the duration of the option is one year.

16.2 For the data of Problem 16.1, answer the same questions about a *put* option on the stock.

16.3 The current price of HAL computer stock is \$280. Its closing price during each of the next twelve months is given in Table 16.4.

Table 16.4

Month	Closing Price
1	\$290.00
2	\$300.00
3	\$282.00
4	\$286.50
5	\$291.40
6	\$308.00
7	\$298.32
8	\$304.50
9	\$308.23
10	\$310.40
11	\$308.21
12	\$309.34

Use this data to estimate the annual mean and standard deviation of the return on HAL stock.

16.4 Suppose that you have analyzed daily price data on Macburger stock and found that μ_{daily} = .001 and σ_{daily} =.00970. Assuming the stock market is open 260 days per year, estimate the mean and standard deviation of the annual return on Macburger stock.

Group B

16.5 If you own a stock, buying a put option on the stock will greatly reduce your risk. This is the idea behind **portfolio insurance**. To illustrate, consider a stock (Trumpco) that currently sells for \$56 and has σ = 30%. Assume the risk-free rate is 8% and you estimate μ = 12%.

a You own one share of Trumpco. Use simulation to estimate the probability distribution of the percentage return earned on this stock during a one-year period.

b Now suppose you also buy a put option (for $2.38) on Trumpco. The option has an exercise price of $50 and a one-year expiration date. Use simulation to estimate the probability distribution of the percentage return on your portfolio over a one-year period. Can you see why this strategy is called a portfolio insurance strategy?

c Use simulation to show that the put option should, indeed, sell for $2.38.

16.6 For the data in Problem 16.5, the following is an example of a **butterfly spread**: Buy two calls with an exercise price of $50; sell one call with an exercise price of $40 and one call with an exercise price of $60. Simulate the cash flows from this portfolio.

16.7 Cryco stock currently sells for $69. The annual growth rate of the stock is 15%, and the stock's annual volatility is 35%. The risk-free rate is currently 5%. You have bought a six-month European put option on this stock with an exercise price of $70.

a Use @RISK to value this option.

b Use @RISK to analyze the distribution of percentage returns (for a six-month horizon) for the following portfolios:

Portfolio 1: Own one share of Cryco.

Portfolio 2: Own one share of Cryco and buy the put described in part (a).

Which portfolio has the largest expected return? Explain why Portfolio 2 is known as portfolio insurance.

16.8 Consider a stock currently selling for $40. Consider a European call option with exercise price $41 and duration 3 months. Suppose the stock price three months from now is certain to be either $42 or $38 and the risk-free rate is 12%.

a Suppose we are long one share of the stock and short four call options on the stock. Show that whether the stock goes up or down, the value of our portfolio remains the same.

b Since the portfolio in part (a) is riskless, it must earn the risk-free rate over a three-month period. Use this fact to determine a fair price for the call option.

c Does the price for the call option obtained in part (b) depend on the mean growth rate of the stock?

References

Black, F., and M. Scholes. 1973. "The Pricing of Options and Corporate Liabilities." *Journal of Political Economy* 81: 637–654.

Cox, J., S. Ross, and M. Rubenstein. 1979. "Option Pricing: A Simplified Approach." *Journal of Financial Economics.* 7: 229–263.

CHAPTER 17

Pricing Path-Dependent and Exotic options

There are many options whose payoff depends on the price of the underlying security at times other than the option's expiration time. These are called **path-dependent** or **exotic options**.

Here are three examples of exotic options:

Asian Option: The payoff from an Asian option depends on the average price of the underlying security over the lifetime of the option. Let the exercise price of an Asian option be X and let S_{avg} be the average price of the stock over the lifetime of the option. The payoff from an Asian call option is

$$\text{Max}(S_{avg} - X, 0) \tag{17.1}$$

The payoff from an Asian put option is $\text{Max}(X - S_{avg}, 0)$.

Lookback Option: Let S_L be the lowest price reached during the life of an option and S_H be the highest price reached during the life of an option. Then the payoff from a lookback call is $S_T - S_L$ and the payback from a lookback put is $S_H - S_T$. S_T is the stock price on the option's termination date.

Knockout Option: A **down-and-out knockout option** is "knocked out" or yields no payoff if the price of the stock reaches a barrier H during the life of the option. A **down-and-in option** yields a payoff only if the price of the stock reaches a level H during the life of the option.

To price an exotic option, we simply take the expected discounted value of the payoffs from the option, again assuming that the stock grows at the risk-free rate. In generating the path of stock prices, we again use (16.1). The trick is to make sure that the stock price at, say, Time t is generated from (16.1) with the price at Time $t - 1$ playing the role of P_0. If we fail to do this we might find a very low price of the stock on Day $t - 1$ followed by a very high price on Day t. This would not make sense!

We now give an example of pricing a path-dependent option.

Example 17.1

Consider a 52-week Asian call option. Today's price is $100, and the exercise price is $110. The annual volatility for the stock is 30% and the risk-free rate is 9%. The payoff to the option depends on the average price of the stock at weekly intervals (beginning with today). What is a fair price for this option?

Solution Our work is in the file Asian.xls or Asian.wk4. See Figure 17.1 for the completed spreadsheet. We proceeded as follows:

Step 1 In the cell range B4:B8 we entered the relevant information about the option.

Step 2 In cell B13 we recopy the initial stock price with the formula

```
Excel:  =B4

Lotus:  +B4
```

Step 3 In cell B14 we generate the stock price one week into the option with the formula

```
Excel:  =B13*EXP(($B$7  -.5*$B$6^2)*(1/52)  +
$B$6*RISKNORMAL(0,1)*(1/52)^.5)

Lotus:  +B13*EXP(($B$7-.5*$B$6^2)*(1/52)  +
$B$6*RISKNORMAL(0,1)*(1/52)^.5)
```

Note that we are assuming the stock grows at the risk-free rate.

Copying this formula to the cell range B15:B65 creates weekly stock prices for the entire life of the option.

Step 4 In C11 we compute the average weekly price of the option with the formula

```
Excel:  =AVERAGE(B13:B65)

Lotus:  @AVG(B13:B65)
```

Step 5 Recall that the payoff from an Asian call option is given in (17.1). To operationalize this formula we enter in cell C10 the formula

	A	B	C	D	E	F
1		**Pricing an Asian Option**				
2		**Figure 17.1**				
3						
4	Initial Price	$ 100.00				
5	Exercise Price	$ 110.00				
6	Annual Volatili	0.3				
7	Risk-Free Rate	0.09				
8	Duration(years	1				
9		Discounted Cashflow value	12.39833			
10		Cashflow from option at terminat	$ 13.57			
11		Average Price	$ 123.57			
12	Week	Stock Price				
13	0	$ 100.00				
14	1	97.80280012				
15	2	101.1029158			Cell	C9
16	3	104.2607992			Minimum =	0
17	4	109.2577469			Maximum =	69.197
18	5	107.6876897			Mean =	4.5608
19	6	108.4215725			Std Deviation =	9.3374
20	7	119.9786771			Variance =	87.188
21	8	122.264485			Skewness =	2.7371
22	9	119.2913229			Kurtosis =	12.039
23	10	109.2581395			Errors Calculated	0
24	11	119.0375459			Mode =	0
25	12	119.0949637			5% Perc =	0
26	13	124.5751436			10% Perc =	0
27	14	120.9086151			15% Perc =	0
28	15	134.2864366			20% Perc =	0
29	16	142.8643753			25% Perc =	0
30	17	137.2593422			30% Perc =	0
31	18	132.0874098			35% Perc =	0
32	19	132.9935041			40% Perc =	0
33	20	126.7608216			45% Perc =	0
34	21	129.0868524			50% Perc =	0
35	22	129.7187734			55% Perc =	0
36	23	124.8836694			60% Perc =	0
37	24	122.838921			65% Perc =	0
38	25	123.9792454			70% Perc =	2.2171
39	26	121.3938368			75% Perc =	5.0673
40	27	117.6820776			80% Perc =	7.9532
41	28	132.0550517			85% Perc =	10.81
42	29	136.4158499			90% Perc =	17.221
43	30	128.4016639			95% Perc =	25.757
44	31	129.9102612				
45	32	122.8509133				
46	33	120.2977163				
47	34	115.9120746				
48	35	116.2930086				
49	36	121.0954066				
50	37	120.0932804				
51	38	118.0879897				
52	39	113.6093237				
53	40	122.9975995				
54	41	122.9165866				
55	42	123.7235274				
56	43	132.5940127				
57	44	139.7927616				
58	45	143.1904249				
59	46	142.9539132				
60	47	138.9241532				
61	48	142.8730229				
62	49	136.6348607				
63	50	132.6538952				
64	51	120.9066231				
65	52	115.0330199				

Figure 17.1 Asian Option

```
Excel:  =MAX(C11-B5,0)

Lotus:  @MAX(C11-B5,0)
```

Step 6 The discounted cash flow from the option is computed in cell C9 with the formula

```
Excel:  =EXP(-B7*B8)*C10
```

Step 7 We selected cell C9 as our output cell and ran 400 iterations.

@RISK indicated the Asian option should sell for $4.56. The actual price turns out to be $4.68. The high standard deviation (9.34) for the option's discounted payoff indicates that more iterations are needed to accurately price the option.

Problems

Group A

17.1 A **knockout call option** loses all value at the instant the price of the stock drops below a given "knockout level." Determine a fair price for a knockout call option in the following situation:

Current Stock Price: $20. | Knockout Price: $19.50.

Exercise Price: $21. | Annual Volatility: 40%.

Risk-free Rate: 10%. | Mean Growth Rate of Stock: 12%.

Duration of Option: 1 month = 21 days; assume 250 days = 1 year.

17.2 An increase in volatility will surely increase the price of a call or put option, but an increase in volatility may decrease the value of a knockout option. Why?

17.3 If you wanted to hedge holdings in a stock, what advantage would a knockout put have over a regular put?

17.4 Consider the stock described in Problem 17.1.

a Price a lookback call option with a duration of 6 months.

b Price a lookback put option with a duration of 6 months.

17.5 For the Asian option of Example 17.1 determine how a change in volatility affects the option's value.

17.6 Suppose that 20 weeks before the averaging period for the Asian option of Example 17.1 begins, the stock sells for $100. At this time what would be a fair price for the option?

17.7 Suppose that 20 weeks into the averaging period (i.e., in row 33) for the Asian option of Example 17.1, the stock is selling for $100 and the average price of the stock during the first 21 averaging points (initial price plus first 20 weeks) has been $100. At this point in time what is a fair price for the option?

CHAPTER 18

Using Immunization to Manage Interest Rate Risk

We all are familiar with the savings and loan debacle of the 1980s, which cost our nation billions and billions (as Carl Sagan would say!) of dollars. A major cause of the S and L crisis was the inability of S and Ls to manage interest rate risk. To illustrate interest rate risk look at Figure 18.1 (file Sandl.wk4 or Sandl.xls). Suppose this represents the balance sheet of Keating Trust. All figures are in millions of dollars. The bank has a liability of $560 million due at the beginning of Year 1 and a liability of $400 million due at the beginning of Year 2. Keating Trust has its assets tied up in home mortgage loans. They expect to receive payments of $100 million at the beginning of each of the next 20 years. At present the interest rate is 10%. We proceed as follows:

Step 1 In cell E3 we enter the interest rate 10%.

Step 2 In cell C34 we compute the NPV of all assets with the formula

```
Excel:  =B8+NPV(E3,B9:B27)
Lotus:  B8+@NPV(E3,B9..B27)
```

We find the NPV of the assets to be $936.49 million. Similarly, in cell C37 we compute the NPV of the bank's liabilities to be $923.64 million. Keating appears to be in good shape! The problem is that interest rates change over time. To model the random nature of interest rates we proceed as follows:

Step 1 Historically, the change in interest rates from one year to the next has a mean of 0 and a standard deviation of 2.7%, which we entered in cell E4.

	A	B	C	D	E	F	G
1	**Interest rate risk**		**Figure 18.1**				
2							
3			Current Interest Rate		0.10		
4	Annual Standard Deviation of interest rates				0.027		
5			Npv of assets		Npv of liabilities		Random
6	Year		with random		with random	Interest	discount
7		Asset	discount factor	Liability	discount facto	rate	factor
8	1	100.00	100.00	560.00	560.00	0.10	1.00
9	2	100.00	90.91	400.00	363.64	0.08	0.91
10	3	100.00	83.84			0.09	0.84
11	4	100.00	76.86			0.12	0.77
12	5	100.00	68.51			0.12	0.69
13	6	100.00	61.17			0.10	0.61
14	7	100.00	55.55			0.10	0.56
15	8	100.00	50.45			0.11	0.50
16	9	100.00	45.55			0.07	0.46
17	10	100.00	42.43			0.04	0.42
18	11	100.00	40.62			0.07	0.41
19	12	100.00	37.82			0.07	0.38
20	13	100.00	35.41			0.04	0.35
21	14	100.00	33.89			0.05	0.34
22	15	100.00	32.27			0.00	0.32
23	16	100.00	32.30			0.01	0.32
24	17	100.00	31.95			0.03	0.32
25	18	100.00	31.04			0.04	0.31
26	19	100.00	29.84			0.05	0.30
27	20	100.00	28.51			0.06	0.29
28		Asset		Liability			
29		Npv	1008.93	Npv	923.64		
30							
31	Asset-Liability Npv		85.30		Variable Type		
32					Name	Asset-Liability Npv/dis	
33					Description	Output	
34	Npv of assets		$936.49		Cell	C31	
35	at 10%				Minimum =	-366.53	
36	Npv of liabilities				Maximum =	4382.53	
37	at 10%		$923.64		Mean =	129.14	
38					Std Deviation	434.92	
39					Variance =	189154.40	
40					Skewness =	5.11	
41					Kurtosis =	46.92	
42					Errors Calcula	0.00	
43					Percentile Values		
44					5% Perc =	-248.09	
45					10% Perc =	-195.56	
46					15% Perc =	-163.24	
47					20% Perc =	-132.25	
48					25% Perc =	-115.68	
49					30% Perc =	-94.90	
50					35% Perc =	-65.60	
51					40% Perc =	-41.58	
52					45% Perc =	3.31	
53					50% Perc =	31.32	
54					55% Perc =	65.43	
55					60% Perc =	92.22	
56					65% Perc =	146.59	
57					70% Perc =	178.28	
58					75% Perc =	231.92	
59					80% Perc =	275.87	
60					85% Perc =	432.27	
61					90% Perc =	577.04	
62					95% Perc =	874.89	

Figure 18.1 Keating Five and Dime

Step 2 We model the movement of interest rates in column F. In cell F8 we enter the current interest rate with the formula

```
Excel: =E3

Lotus: +E3
```

Step 3 To generate the Year 2 interest rate we enter into cell F9

```
Excel: =F8+RISKNORMAL(0,E$4)

Lotus: +F8+@RISKNORMAL(0,E$4)
```

Copying this formula to the range F10:F27 generates 20 years of interest rates.

We are interested in the probability distribution of the NPV of Assets – NPV of Liabilities for the bank. To compute the NPV of the asset and liability payment streams we need to use the interest rates generated in column F to compute the discount factor for each year. Let i_t be the Year t interest rate. For Year 1 the discount factor is 1 (because payments are assumed to occur at the beginning of each year, not the end). Then for $t > 1$ the discount factor for Year t is given by

$$\text{Discount Factor for Year } t = \frac{1}{(1+i_1)(1+i_2)\ldots(1+i_{t-1})} \tag{18.1}$$

Note that

$$\text{Year } t \text{ Discount Factor} = \frac{(\text{Year } t-1 \text{ Discount Factor})}{(1+i_{t-1})}$$

Step 4 To generate the discount factors we begin by entering a 1 in cell G8. Next we enter in cell G9

```
Excel: =G8/(1+F8)

Lotus: +G8/(1+F8)
```

Copying this formula to the range G9:G27 operationalizes (18.1) and creates the discount factor for each year.

Step 5 To compute the NPV of the asset payment stream for each year begin by entering in cell C8

```
Excel: =B8*G8

Lotus: +B8*G8
```

Copying this formula to the range C9:C27 generates for each year the product and the asset payment and the discount factor.

Step 6 Entering the formula in cell C29

```
Excel =SUM(C8:C27)

Lotus: @SUM(C8..C27)
```

computes the NPV of the asset stream. When all interest rates are 10% this yields $936.49.

Step 7 Following a similar procedure in column E we generate the NPV of the liability payment stream in cell E29. To compute the NPV of Assets – NPV of Liabilities we enter in cell C31

```
Excel: =C29-E29

Lotus: +C29-E29
```

Step 8 We now choose cell C31 as our output cell and use @RISK to simulate the worksheet 200 times.

From Figure 18.1 you see that the standard deviation of the NPV of Assets – NPV of Liabilities was $434.9 million! There was a 30% chance that the bank would be more than $94 million in the hole! Clearly there is too much risk here.

The reason Keating is exposed to such high interest rate risk is that many of their assets pay off in the distant future. The discount factor for payments in the distant future is much more sensitive to interest rate variation than the discount factor for payments in the near future. One way to "immunize" against interest rate risk is to match (using the concepts of duration and convexity) the timing of your asset stream to your liability stream. Before discussing immunization, we will define the concepts of duration and convexity of a payment stream.

18.1 Duration

Suppose that at the end of Year t ($t = 1, 2, \ldots, N$) you receive $\$c_t$. Given an Interest Rate i, the **duration** of this payment stream is defined by

$$\frac{\displaystyle\sum_{t=1}^{t=N} \frac{tc_t}{(1+i)^t}}{\displaystyle\sum_{t=1}^{t=N} \frac{c_t}{(1+i)^t}}$$

Intutitively, the duration of a payment stream is the expected time at which a randomly chosen dollar of NPV is received. The higher the duration of a payment stream, the more sensitive the value of the payment stream is to changes in interest rates. It can be shown if the duration of assets equals D and the interest rate at all-points in time increases from its current value of r by a small amount Δr, then

$$\text{Change in NPV of Assets} \doteq \frac{-D * \Delta r(\text{NPV of Assets})}{(1 + r)}$$

18.2 Convexity

Suppose that at the end of Year t ($t = 1, 2, \ldots, N$) you receive $\$c_t$. Given an Interest Rate i, the **convexity** of this payment stream is defined by

$$\frac{\sum_{t=1}^{t=N} \frac{t(t+1)c_t}{(1+i)^t}}{\sum_{t=1}^{t=N} \frac{c_t}{(1+i)^t}}$$

The convexity of a payment stream is approximately the rate of change of the duration of the portfolio with respect to a change in interest rates. A large value of convexity means that a small change in the current interest rate will result in a large change in a portfolio's duration, while a small value of convexity means that a small change in the current interest rate will result in a small change in the portfolio's duration.

18.3 Immunization Against Interest Rate Risk

The key idea of immunization can be stated thus: *If your liabilities and assets have the same values of duration and convexity, then you will be subject to very little interest rate risk.*

To illustrate how immunization (often called **duration matching**) works consider the following example.

Example 18.1

Your bank has the obligation to payout $163 at the end of each of the next 15 years. With the current interest rate of 10%, this obligation has an NPV of $1,239.79. There are three bonds that you are considering for investment:

Bond 1: A five-year bond paying $100 at the end of Years 1–4 and $1100 at the end of Year 5.

Bond 2: A ten-year bond paying $100 at the end of Years 1–9 and $1100 at the end of Year 10.

Bond 3: A fifteen-year bond paying $100 at the end of Years 1–14 and $1100 at the end of Year 15.

How should the bank invest to best immunize themselves against interest rate risk?

Solution Spreadsheet computations (illustrated in Figure 18.2 and files Convexit.wk4 and Convexit.xls) yield the durations and convexities given in Table 18.1.

Table 18.1 Durations and Convexities

	Duration	Convexity
Liabilities	6.28	62.60
Bond 1	4.17	23.44
Bond 2	6.76	63.88
Bond 3	8.37	105.07

Let n_i be the number of Bond i that should be purchased. Each bond's payments have an NPV of \$1000, so we assume the price of each bond is \$1000. Since our assets need to have an NPV of \$1,239.79, we must have

$$1000n_1 + 1000n_2 + 1000n_3 = 1{,}239.79 \quad \textbf{(18.2)}$$

Suppose M assets are available and an amount p_j has been spent to purchase asset j. Then the duration of the portfolio of assets may be approximated by

$$\frac{\sum_{j=1}^{j=M} p_j(\text{Duration of Asset } j)}{\sum_{j=1}^{j=M} p_j}$$

and the convexity of the portfolio of assets may be approximated by

$$\frac{\sum_{j=1}^{j=M} p_j(\text{Convexity of Asset } j)}{\sum_{j=1}^{j=M} p_j}$$

Thus to match the duration of the asset portfolio to the duration of the liabilities, the n_j must satisfy

$$\frac{4.17*(1000n_1)+6.76*(1000*n_2)+8.37*(1000n_3)}{1{,}239.79} = 6.28 \quad \textbf{(18.3)}$$

Finally, to match the convexity of the asset portfolio to the convexity of the liabilities, the n_j must satisfy

$$\frac{23.44*(1000n_1)+63.88*(1000*n_2)+105.07*(1000n_3)}{1{,}239.79} = 62.60 \quad \textbf{(18.4)}$$

	A	B	C	D	E	F	G
1	YEAR	LIABILITY	ASSET3	T*CT/PRICE	T*CT/PRICE	T*(T+1)*CT	T*(T+1)*CT
2				LIABILITY	ASSET3	/PRICE	/PRICE
3						LIABILITY	ASSET3
4	1	163	100	0.1314737777	0.1	0.262947554	0.2
5	2	163	100	0.262947554	0.2	0.788842661	0.6
6	3	163	100	0.3944421331	0.3	1.577685323	1.2
7	4	163	100	0.5258895108	0.4	2.629475538	2
8	5	163	100	0.6573688884	0.5	3.944213307	3
9	6	163	100	0.7888842661	0.6	5.521898629	4.2
10	7	163	100	0.9203316438	0.7	7.362531506	5.6
11	8	163	100	1.0517790215	0.8	9.466111936	7.2
12	9	163	100	1.1832263992	0.9	11.83263992	9
13	10	163	100	1.3147377769	1	14.46211546	11
14	11	163	100	1.4462211546	1.1	17.35453855	13.2
15	12	163	100	1.5776885323	1.2	20.50990919	15.6
16	13	163	100	1.7091591	1.3	23.92822739	18.2
17	14	163	100	1.8406632876	1.4	27.60949315	21
18	15	163	1100	1.9721066653	16.5	31.55370645	264
19	16	1239.791		6.2789333467	8.366687457	62.59947174	105.0677478

Figure 18.2 Determination of Convexity and Duration

	A	B	C	D	E	F	G	H	I
1	**Computation of**								
2	**Immunized**								
3	**Portfolio**								
4			Number	Number	Number				
5			Bond1	Bond2	Bond3				
6			0.528151	0.231633	0.480007				
7	Dollars in Each Bond		528.1507	231.6325	480.0068				Targets
8	Npv of Each Bond		1000	1000	1000	Portfolio Npv	1239.79	=	1239.79
9	Duration of Each bond		4.17	6.76	8.37	Portfolio Duration	6.28	=	6.28
10	Convexity of each bond		23.44	63.88	105.07	Portfolio Convexity	62.6	=	62.6

Figure 18.3 Duration and Convexity Matching with the Solver

Solving (18.2)-(18.4) simultaneously (or using the Excel Solver or What's Best, see files Bondwt.xls or Bondwt.wk4 and Figure 18.3) we find that n_1 = .52815, n_2 = .23164, and n_3 = .48001.

To use a spreadsheet solver to determine the number of each bond purchased that yields the appropriate NPV, duration, and convexity we proceed as follows:

Step 1 Begin by entering in cells C6:E6 trial values of the number of each bond purchased.

Step 2 In the cell range C7:E7 we compute the number of dollars invested in each bond (this will be used to determine the duration and convexity of our bond portfolio). The number of dollars invested in Bond 1 is computed in cell C7 with the formula

```
Excel:  =1000*C6

Lotus:  1000*C6
```

Copying this formula to the range D7:E7 computes the number of dollars invested in each bond.

Step 3 We enter the NPV of Bonds 1–3, respectively, in the cell range C8:E8. Then in cell G8 we compute the NPV of our bond portfolio with the formula

```
Excel:  =SUMPRODUCT(C6:E6,C8:E8)

Lotus:  @SUMPRODUCT(C6..E6,C8..E8)
```

We will set this equal to the required NPV of $1239.79, which is entered in cell I8.

Step 4 We enter the duration of Bonds 1–3, respectively, in the cell range C9:E9.

Step 5 In cell G9 we compute the duration of our bond portfolio with the formula

```
Excel:  =SUMPRODUCT(C$7:E$7,C9:E9)
/I$8

Lotus:  @SUMPRODUCT(C$7..E$7,C9..E9)
/I$8
```

We will set this equal to the required duration of 6.28.

Step 6 After entering the convexity of each bond in the cell range C10:E10, we compute the duration of the portfolio by copying the formula in G9 to the cell G10. We will set the convexity of the portfolio equal to the target of 62.6, which is entered in cell I10.

Solving for the Bond Weights with What's Best

Step 1 Choose the cell range B6..D6 as your adjustable cells.

Step 2 Use the Constraint menu to constrain the range G8..G10 to equal the range I8..I10.

Step 3 Hitting Solve yields the solution in Figure 18.3. Note that we do not have a best cell in this example!

Solving for the Bond Weights with the Excel Solver

Step 1 Choose the cell range C6:E6 as your changing cells.

Step 2 Add the constraints $G$8:$G$10=$I$8:$I$10.

Step 3 Hit Solve and choose the linear model option to obtain the solution in Figure 18.3. Note that we do not need a best cell.

Remark

Matching durations and convexity may require the "purchase" of a negative number of one or more bonds. This is OK. A purchase of a negative number of a bond is equivalent to the bank selling a bond. This means that the bank pays out rather than receives the coupons.

In Figure 18.4 (file Duration.wk4 or file Duration.xls) we set up a spreadsheet to simulate the sensitivity of NPV of Assets – NPV of Liabilities to interest rate risk. We proceed as follows:

Step 1 In column B enter the liability cash flows for each year (see Figure 18.4a).

Step 2 In columns C–E enter the cash flows for each year for one unit of each bond.

Step 3 In columns F–H enter for each year the payout from our holdings of each bond. These depend, of course, on our holdings of each bond, which are given in the cell range D3:D5.

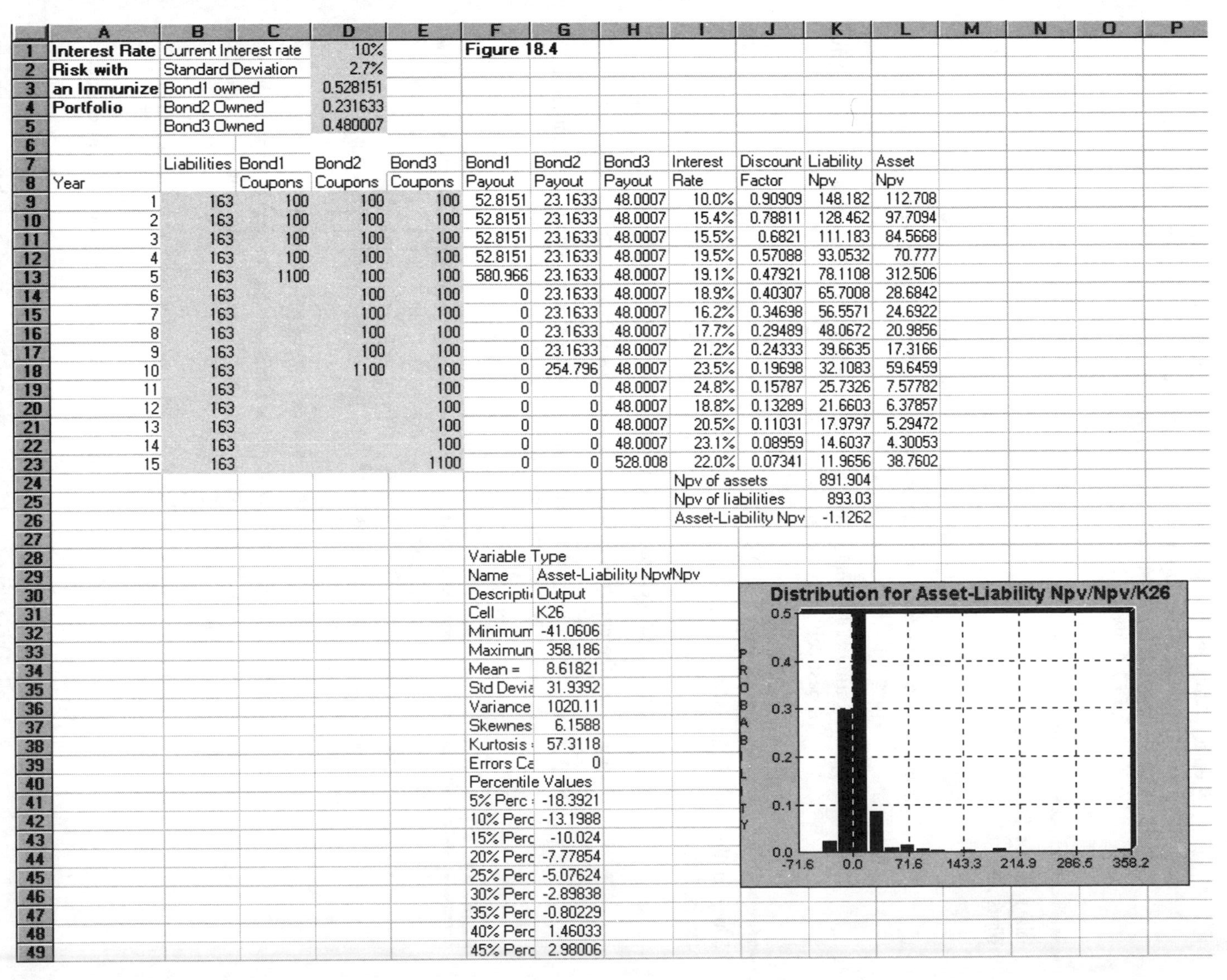

	A	B	C	D	E	F	G	H	I	J	K	L
1	**Interest Rate**	Current Interest rate		10%		**Figure 18.4**						
2	**Risk with**	Standard Deviation		2.7%								
3	**an Immunize**	Bond1 owned		0.528151								
4	**Portfolio**	Bond2 Owned		0.231633								
5		Bond3 Owned		0.480007								
6												
7		Liabilities	Bond1	Bond2	Bond3	Bond1	Bond2	Bond3	Interest	Discount	Liability	Asset
8	Year		Coupons	Coupons	Coupons	Payout	Payout	Payout	Rate	Factor	Npv	Npv
9	1	163	100	100	100	52.8151	23.1633	48.0007	10.0%	0.90909	148.182	112.708
10	2	163	100	100	100	52.8151	23.1633	48.0007	15.4%	0.78811	128.462	97.7094
11	3	163	100	100	100	52.8151	23.1633	48.0007	15.5%	0.6821	111.183	84.5668
12	4	163	100	100	100	52.8151	23.1633	48.0007	19.5%	0.57088	93.0532	70.777
13	5	163	1100	100	100	580.966	23.1633	48.0007	19.1%	0.47921	78.1108	312.506
14	6	163		100	100	0	23.1633	48.0007	18.9%	0.40307	65.7008	28.6842
15	7	163		100	100	0	23.1633	48.0007	16.2%	0.34698	56.5571	24.6922
16	8	163		100	100	0	23.1633	48.0007	17.7%	0.29489	48.0672	20.9856
17	9	163		100	100	0	23.1633	48.0007	21.2%	0.24333	39.6635	17.3166
18	10	163		1100	100	0	254.796	48.0007	23.5%	0.19698	32.1083	59.6459
19	11	163			100	0	0	48.0007	24.8%	0.15787	25.7326	7.57782
20	12	163			100	0	0	48.0007	18.8%	0.13289	21.6603	6.37857
21	13	163			100	0	0	48.0007	20.5%	0.11031	17.9797	5.29472
22	14	163			100	0	0	48.0007	23.1%	0.08959	14.6037	4.30053
23	15	163			1100	0	0	528.008	22.0%	0.07341	11.9656	38.7602
24									Npv of assets		891.904	
25									Npv of liabilities		893.03	
26									Asset-Liability Npv		-1.1262	
27												
28						Variable Type						
29						Name	Asset-Liability NpvNpv					
30						Descripti	Output					
31						Cell	K26					
32						Minimum	-41.0606					
33						Maximun	358.186					
34						Mean =	8.61821					
35						Std Devia	31.9392					
36						Variance	1020.11					
37						Skewnes	6.1588					
38						Kurtosis	57.3118					
39						Errors Ca	0					
40						Percentile Values						
41						5% Perc	-18.3921					
42						10% Perc	-13.1988					
43						15% Perc	-10.024					
44						20% Perc	-7.77854					
45						25% Perc	-5.07624					
46						30% Perc	-2.89838					
47						35% Perc	-0.80229					
48						40% Perc	1.46033					
49						45% Perc	2.98006					

Figure 18.4 Computation of Interest Rate Risk for Immunized Portfolio

	A	B	C	D	E	F	G	H	I	J	K	L
1	**Interest Rate**	Current Interest rate		10%		**Figure 18.4**						
2	**Risk with**	Standard Deviation		2.7%								
3	**an Immunized**	Bond1 owned		0.52815								
4	**Portfolio**	Bond2 Owned		0.23163								
5		Bond3 Owned		0.48001								
6												
7		Liabilities	Bond1	Bond2	Bond3	Bond1	Bond2	Bond3	Interest	Discount	Liability	Asset
8	Year		Coupons	Coupons	Coupons	Payout	Payout	Payout	Rate	Factor	Npv	Npv
9	1	163	100	100	100	52.8151	23.1633	48.0007	10.0%	0.90909	148.182	112.708
10	2	163	100	100	100	52.8151	23.1633	48.0007	15.4%	0.78811	128.462	97.7094
11	3	163	100	100	100	52.8151	23.1633	48.0007	15.5%	0.6821	111.183	84.5668
12	4	163	100	100	100	52.8151	23.1633	48.0007	19.5%	0.57088	93.0532	70.777
13	5	163	1100	100	100	580.966	23.1633	48.0007	19.1%	0.47921	78.1108	312.506
14	6	163		100	100	0	23.1633	48.0007	18.9%	0.40307	65.7008	28.6842
15	7	163		100	100	0	23.1633	48.0007	16.2%	0.34698	56.5571	24.6922
16	8	163		100	100	0	23.1633	48.0007	17.7%	0.29489	48.0672	20.9856
17	9	163		100	100	0	23.1633	48.0007	21.2%	0.24333	39.6635	17.3166
18	10	163		1100	100	0	254.796	48.0007	23.5%	0.19698	32.1083	59.6459
19	11	163			100	0	0	48.0007	24.8%	0.15787	25.7326	7.57782
20	12	163			100	0	0	48.0007	18.8%	0.13289	21.6603	6.37857
21	13	163			100	0	0	48.0007	20.5%	0.11031	17.9797	5.29472
22	14	163			100	0	0	48.0007	23.1%	0.08959	14.6037	4.30053
23	15	163			1100	0	0	528.008	22.0%	0.07341	11.9656	38.7602
24									Npv of assets		891.904	
25									Npv of liabilities		893.03	
26									Asset-Liability Npv		-1.1262	

Figure 18.4a

Figure 18.4b

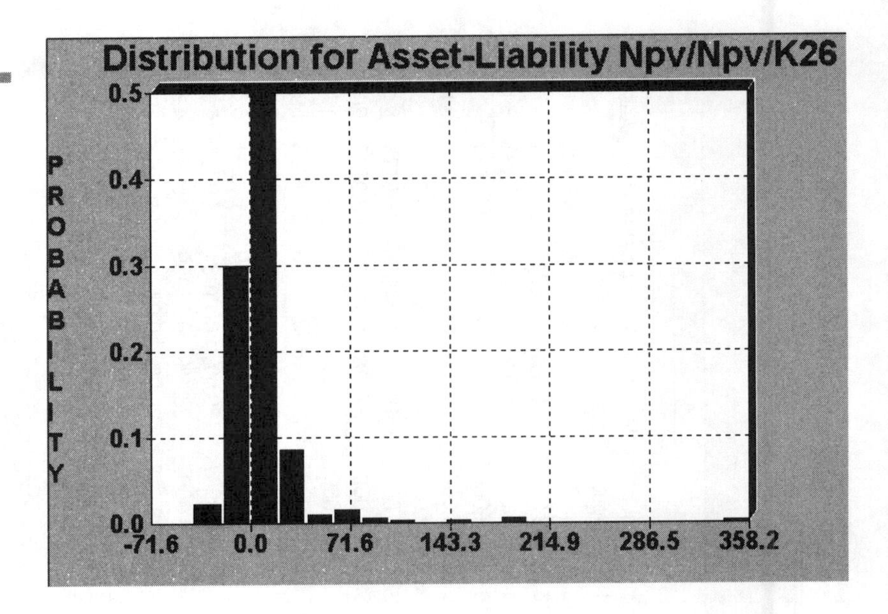

Step 4 Column I generates the interest rates for each year in the same manner as in Figure 18.1.

Step 5 Column J computes the discount factor for each year. Note that each payment is received at the end of each year so in cell J9 we do not enter a 1, we enter

```
Excel: =1/(1+I9)
Lotus: 1/(1+I9)
```

Step 6 To generate the Year 2 discount factor we enter in cell J10 the formula

```
Excel: =J9/(1+I10)
Lotus: +J9/(1+I10)
```

Copying from cell J10 to the cell range J11:J23 generates the discount factors for Years 3–15.

Step 7 In column K we compute the contribution to NPV of liability payments for each year. Total NPV of liabilities is computed in K25.

Step 8 In column L we compute the contribution to NPV of assets for each year. Total NPV of assets is computed in cell K24. In cell K26 we compute NPV of Assets – NPV of Liabilities.

Figure 18.4c

Variable Type			
Name	Asset-Liability Npv/Npv		
Descriptioı	Output		
Cell	K26		
Minimum =	-41.0606		
Maximum	358.1861		
Mean =	8.618212		
Std Deviati	31.93919		
Variance =	1020.112		
Skewness	6.158798		
Kurtosis =	57.31179		
Errors Cal	0		
Percentile Values			
5% Perc =	-18.3921		
10% Perc	-13.1988		
15% Perc	-10.024		
20% Perc	-7.77854		
25% Perc	-5.07624		
30% Perc	-2.89838		
35% Perc	-0.80229		
40% Perc	1.460328		
45% Perc	2.980064		
50% Perc	4.957131		
55% Perc	7.631479		
60% Perc	8.509398		
65% Perc	10.11333		
70% Perc	12.19704		
75% Perc	13.71665		
80% Perc	15.40641		
85% Perc	18.28467		
90% Perc	21.98977		
95% Perc	36.05933		

Step 9 We now used @RISK to simulate this spreadsheet 200 times. Our output cell was K26. Observe that the standard deviation of the values assumed by cell K26 was 31.9 (see Figures 18.4b and 18.4c). There is only a 15% chance that we will be more than $10.02 (out of $1,239.79!) in the hole. Clearly, we have done a good job of immunizing ourselves against interest rate risk!!

18.4 Better Models for Interest Rate Risk

In reality the intervention of the Federal Reserve Bank prevents interest rates from ever reaching very high levels. Our current interest rate model allows interest rates of 30% or 40% per year. This is clearly unrealistic. Our model of interest rates also allows for negative interest rates. This is also unrealistic. A better model for interest rates was proposed by Vasicek (1977). In his **mean reversion model** interest rates revert to a given level (say 10%). To make this happen suppose that the mean growth in the interest rate during a year is a*(0.10 – Current Interest Rate), where a is a constant. This forces the mean change in interest rates to be positive when they are lower than 10% and forces the mean change in interest rates to be negative when interest rates are higher than 10%. To operationalize this model (with $a = 0.2$) in the Keating Five and Dime example we would enter the value of a in cell E2 of Figure 18.1 and the level to which interest rates revert (10%) in cell E1. Then we would generate Year 2 interest rates in cell F9 with the formula

```
Excel:  =F8+RISKNORMAL(E$2*(E$1-F8),E$4)

Lotus:  +F8+@RISKNORMAL(E$2*(E$1-F8),E$4)
```

Copying this formula to the cell range F10:F27 generates 20 years of interest rates.

Vasicek's model can yield negative interest rates. The Cox-Ingersoll-Ross (1985) model uses the same formula as Vasicek for the mean change in interest rates but the Cox-Ingersoll-Ross model prevents negative interest rates from occuring by setting the standard deviation of the change in interest rates equal to b*(current interest rate)$^{1/2}$. Here b is a constant ($b = .09$ will yield the same variability in the change in interest rates as our current model when interest rates are 10%). When interest rates approach 0 then (current interest rate)$^{1/2}$ approaches 0, and the variability of the change in interest rates slows down. This makes negative interest rates an unlikely occurrence. See Problem 18.4 for an illustration of the Cox-Ingersoll-Ross model.

Problems

Group A

18.1 At the end of Year 10 you have an obligation to pay $17,908.50. The current interest rate is 10%. You can fund this obligation by purchasing four bonds.

Bond 1: Pays $68 at the end of Years 1–9, and $1068 at the end of Year 10.

Bond 2: Pays $70 at the end of Years 1–14, and $1070 at the end of Year 15.

Bond 3: Pays $60 at the end of Years 1–29 and $1060 at the end of Year 30.

Bond 4: Pays $100 at the end of Years 1–4 and $1100 at the end of Year 5.

a How many of each bond would you purchase?

b Simulate the interest rate risk of this portfolio of bonds and compare it to the interest rate risk if you funded the obligation by simply buying Bond 4.

Hint: A negative number of a bond(s) is allowed. This means that instead of receiving coupon payments, you must make them. This is like "shorting" a bond.

18.2 Use the Vasicek model to simulate the interest rate risk in the S and L example.

18.3 Use the Cox-Ross-Ingersoll model to simulate the interest rate risk in the S and L example.

18.4 Redo Problem 18.1 using the following version of the Cox-Ingersoll-Ross mean reversion model to generate interest rate paths. Assume that

(Change in Interest Rate During Year t + 1) is normally distributed with Mean = .36 * [.062 – (Year t Rate)].

Standard Deviation = .067 * (Year t Interest Rate)^.5.

This model was fit to eight years of monthly interest rate data.

References

Cox, J., J. Ingersoll, and S. Ross. 1985. "A Theory of the Term Structure of Interest Rates." *Econometrica* 53: 385–407.

Vasicek, O. 1977. "An Equilibrium Characterization of the Term Structure." *Journal of Financial Economics* 5: 177–188.

CHAPTER 19

Hedging with Futures

Suppose you have just bought an ounce of gold for $400. The current price of a commodity is called the **spot price** of the commodity. If the price of gold increases, you will make money. If the price of gold decreases, however, you will lose money. To reduce (or **hedge**) this risk you might sell (or **short**) a gold future. Suppose today is January 1, 1995. Selling a gold future (say a one-year future) means that (in effect) one year from now you are paid an amount (called the **futures price**) and in return you must deliver one ounce of gold on January 1, 1996. This obligation may be traded at any time on a futures exchange for the current futures price on a January 1, 1996, future. As time goes on, if the market deems gold more valuable, the value of your gold goes up, but you will lose money on the future contract because you will have to deliver a more valuable commodity. If as time goes on the market deems gold less valuable, the value of your gold decreases, but the value of your future contract increases because you will have to deliver a less valuable commodity. Thus we see that by selling (or shorting) gold futures we can reduce the risk involved in holding gold.

19.1 Hedging with Futures: The Basics

Suppose I need to buy heating oil in November 1994. Also suppose that in June 1994 I buy (**go long**) a December 1994 heating oil future with a price of $1 per barrel. This gives me the right to receive a barrel of heating oil for $1 on December 1, 1994.

Suppose the price of heating oil goes up, and I close out the hedge on November 1, 1994, and buy heating oil. What happens? The cost of buying heating oil goes up (bad) but the value of the future contract goes up (good) because I am receiving a more valuable commodity.

If the price of heating oil drops what happens? The cost of buying heating oil drops (good) but the value of the future drops (bad) because I am receiving a less valuable commodity. Thus whatever happens to the price of heating oil I have hedged my total cost by going long in the futures market.

For example, if I buy oil on November 1, 1994,

Effective Cost of Buying Oil = Cost of Oil on November 1
– (Increase in Market Price of Futures I "Longed" from June 1 to November 1)

Suppose I own an ounce of gold on June 1, 1994. If I sell or short one December 1994 futures hedge at $400 what does this do for me? Shorting the future means I must deliver an ounce of gold (and receive $400) on December 1, 1994.

If gold prices increase, the value of my gold goes up (good) but the value of the future (to me) drops (bad) because I must deliver a more valuable commodity.

If gold prices decrease, then the value of my gold drops (bad), but the value of the future (to me) increases (good) because I must deliver a less valuable commodity. Thus whatever happens to gold prices I have hedged the value of my gold by going short in the futures market.

For example, if a hedge is closed out on November 1, 1994,

(Value of Portfolio on November 1, 1994) =
(Value of Gold on November 1, 1994) – (Increase in Market Price of Futures I Shorted from June 1, 1994 to November 1, 1994)

19.2 Modeling Futures Risk with @RISK

If spot prices and future prices were perfectly correlated, then we could eliminate all risk associated with price changes in the gold market. This is not the case. We can use @RISK to model the risk inherent in a hedge involving futures. Let

S_0 = January 1, 1995, spot price of gold/oz.

S_1 = February 1, 1995, spot price of gold/oz. one month from now.

F_0 = January 1, 1995, futures price for delivering one oz. of gold on January 1, 1996.

F_1 = February 1, 1995, futures price for delivering one oz. of gold on January 1, 1996.

Suppose we sell h ounces in one-year gold futures on January 1, 1995, and buy 1 ounce of gold on January 1, 1995. The return on this "portfolio" has a probability distribution that can be generated by @RISK. Suppose we are given the following information:

Current Gold Spot Price = $400

Average Monthly Increase in Spot Price = $4

Standard Deviation of Monthly Spot Price Changes = $15

Current Futures Price = $440

Standard Deviation of Monthly Futures Price Changes = $18

Correlation Between Monthly Futures Price Changes and Monthly Spot Price Changes = .88

These inputs are entered in the cell range C4:C12 (see Figure 19.1). Note that the spot price one month from now will be normally distributed with a mean = 400 + 4 = $404 and a standard deviation of $15.

The change in the value of your hedged portfolio one month from now is $(S_1 - S_0) - h(F_1 - F_0)$. This is because the gold we have bought for S_0 has increased in value to S_1 while each future contract we have sold (or shorted) has *decreased* in value by an amount $F_1 - F_0$. To model the change in value of this portfolio during a one-month period we need to have @RISK generate the spot and future prices one month from now. Our model can be found in file Gold.wk4 or Gold.xls. We proceed as follows:

Step 1 To generate next month's spot price we enter into cell C14 the following formula:

```
Excel:  =RISKINDEPC("SPOT")+RISKNORMAL(C4+C5,C6)

Lotus:  @RISKINDEPC("SPOT")+@RISKNORMAL(C4+C5,C6)
```

This formula generates observations from a normal random variable with mean 404 and standard deviation 15. The formula also tells @RISK that the spot price is an *independent variable* that can be used to generate future prices.

Step 2 What do we know about the futures price one month from now? Since we expect gold to sell for 400 + 4(12) = $448 a year from now, it is reasonable to expect the futures price to be $448 one year from now. This means that we expect the futures price to increase by (448 – 440)/12 = $0.67 per month (we compute this expected average monthly increase in futures price in cell C8). This means that next month's futures price will be normally distributed with a mean of $440 + $0.67 = $440.67 and a standard deviation of $18. Next month's futures price will also have a correlation of .88 with next month's spot price. To model these relationships we generate next month's futures price in cell C15 with the formula

```
Excel: =RISKDEPC("SPOT",C11)
+RISKNORMAL(C7+C8,C9)

Lotus: @RISKDEPC("SPOT",C11)
+@RISKNORMAL(C7+C8,C9)
```

	A	B	C	D
1	**Hedging Commodity**		**Figure 19.1**	
2	**Risk with Futures**			
3				
4	Current Spot Price		$ 400.00	
5	Mean Spot Increase		$ 4.00	
6	Std. Dev Spot Increase		$ 15.00	
7	Current Futures Price		$ 440.00	
8	Mean Futures increase		$ 0.67	
9	Std. Dev. futures increase		$ 18.00	
10	Expected Spot price in a year		$ 448.00	
11	Correlation between spot and future changes		0.88	
12	Hedge Ratio		0.5	
13				
14	Next month's spot price		389.45928	
15	Next month's future price		468.72179	
16	Change in Portfolio Value		-24.90161	
17				
18		Variable Type		
19		Name	Cell	
20		Description	Output	
21		Cell	C16	
22		Minimum =	-22.75844	
23		Maximum =	33.75468	
24		Mean =	3.67806	
25		Std Deviation =	8.28271	
26		Variance =	68.60329	
27		Skewness =	-0.115972	
28		Kurtosis =	3.284807	
29		Errors Calculated =	0	
30		Percentile Values		
31		5% Perc =	-9.436515	
32		10% Perc =	-7.277012	
33		15% Perc =	-4.660935	
34		20% Perc =	-3.070984	
35		25% Perc =	-1.460314	
36		30% Perc =	-0.423528	
37		35% Perc =	0.5500589	
38		40% Perc =	1.838798	
39		45% Perc =	2.636514	
40		50% Perc =	3.946567	
41		55% Perc =	4.911105	
42		60% Perc =	5.721926	
43		65% Perc =	6.575273	
44		70% Perc =	7.711349	
45		75% Perc =	9.071644	
46		80% Perc =	10.79458	
47		85% Perc =	13.05204	
48		90% Perc =	14.79726	
49		95% Perc =	16.12319	

Figure 19.1 A "Short" Gold Hedge

Step 3 In cell C12 we enter the number of gold futures sold (we chose .5).

Step 4 In cell C16 we now compute the change in the value of the portfolio by entering the formula

```
Excel: =(C14-C4)-C12*(C15-C7)

Lotus: (C14-C4)-C12*(C15-C7)
```

We now use @RISK to simulate the spreadsheet 400 times, with cell C16 as our output cell. We found that the expected change in the value of the hedged portfolio was $3.68 with a standard deviation of $8.28. Using the `@RISKSIMTABLE` function we could have found that a hedge ratio of .7 future contracts sold per ounce of gold purchased would have minimized the variance of the monthly return on the hedged portfolio.

Remarks

1. If you want to minimize the probability that your portfolio would lose, say, more than 5% of its value in a month, you might not choose the hedge ratio that minimizes the variance in the change in your portfolio's value. @RISK could still be used to find the correct hedge ratio, however.

2. Note that if we hold an ounce of gold unhedged, the monthly standard deviation of our portfolio return is $15, while with our hedge we have reduced the standard deviation of the monthly portfolio return to $8.28. We can also see the effect of the hedge by noting from our @RISK output that there is only a 10% chance that our portfolio will lose more than $7.27 in value during a month. If we simply held an ounce of gold, the chance that it would lose at least $7.27 in value can be shown (see Problem 19.4) to be 23%. Thus we see that the hedge reduces the chance of a "bad" outcome.

3. If you do not believe that @RISK has generated correlated spot and future prices, how could you check? Make sure the the Collect Distribution Samples option is on and use the Data option on the results menu to retrieve the actual spot and future prices generated by @RISK. Then use the Clipboard to paste this data into your spreadsheet and use the Excel Correlation function from the Data Analysis Tools menu. You will find the correlation between the spot and future prices generated by @RISK is indeed .88!

Problems

Group A

19.1 Suppose gold currently sells for $300 per ounce. You are going to hedge your holding of 1 oz. of gold by selling six-month futures contracts. You are given the following information:

Current Gold Spot Price = $300

Average Monthly Increase in Spot Price = $5

Standard Deviation of Monthly Spot Price Changes = $15

Current Futures Price = $320

Standard Deviation of Monthly Futures Price Changes = $28

Correlation Between Monthly Futures Price Changes and Monthly Spot Price Changes = .75

How many hedges should you buy to minimize the variability of your hedged portfolio?

19.2 You currently have outstanding the three loans described in Table 19.1.

Table 19.1

Amount of Loan	Interest Rate
$2 million	80% of prime rate
$1 million	2 points above prime rate
$1 million	75% of prime rate

Interest is paid quarterly. During the first quarter of 1996, the prime rate is 14%, and the rate on 90-day T-Bills is 12%. You want to hedge the interest rate risk associated with your next quarterly loan payments by lending out money at the T-Bill rate. Past data indicates that the mean and standard deviation of the quarterly changes in the annual prime and T-Bill rates are as follows:

Table 19.2

	Mean Quarterly Change in Annual Interest Rate	Standard Deviation of Quarterly Change in Annual Interest Rate
Prime	.19%	1.7%
T-Bills	.16%	1.8%

The correlation between the quarterly changes in the prime (annual) rate and the T-Bill (annual) rate is .77.

a Without hedging, simulate the interest rate risk of quarterly payments for the second quarter of 1996.

b Suppose you hedge your risk by lending out money at the next quarter's T-Bill rate. Now simulate (Cost of Interest Payments) – (Loan Payments from T-Bills). Explain how the hedge has decreased your interest rate risk. Use this simulation to determine the amount of T-Bill lending that does the best job of reducing the interest rate risk associated with the second quarter 1996 loan payment.

19.3 Today is December 8, 1995. Currently the S&P 500 index is at 463. Index futures on the S&P for June 1996 delivery are selling for $466. If you buy such a future, you pay $466 (in June 1996) and in June 1996 you receive an amount equal to the June 1996 price of the S&P index. If you sell such a future you receive $466 in June 1996 and must in June 1996 pay out the June 1996 S&P price. Thus if you buy the future you gain if the S&P goes up, while if you sell the future you gain if the S&P drops. Both the index futures and S&P index increase by an average of .5% per month. The standard deviation of the daily changes in the S&P index is 1.34% while the standard deviation of the daily changes in the S&P index futures is 1.82%. The correlation between the S&P index and index futures is .88. Suppose you "own" one unit of the S&P index. How many futures should you "sell" to minimize the variability of your portfolio value?

19.4 Show that during a month the chance that the value of an ounce of gold will drop by $7.27 or more is 23%.

CHAPTER 20

Modeling Market Share

When companies (such as Coca Cola and Pepsi) compete for market share, each firm's market share varies over time. By using advertising and promotions (such as coupons) a firm can increase market share. It is not obvious, however, whether an advertising campaign or promotion will increase profits. As illustrated in the next example, @RISK can be used to model situations in which market share varies over time.

Example 20.1a

Coke and Pepsi are fighting for the cola market. Each week each person in the market buys one case of Coke or Pepsi. If their last purchase was Coke, there is a .90 chance that the person's next purchase will be Coke. If their last purchase was Pepsi, there is a .80 chance that their next purchase will be Pepsi. Currently half of all people purchase each soda. Simulate one year of sales in the cola market and estimate each firm's average market share.

Solution Figure 20.1 (file Cola1.wk4 or Cola1.xls) contains the simulation. To get an idea of what's going on we assume that the total market consists of some reasonably large (but not too large!) number of people—100 will do just fine. We now proceed as follows:

Step 1 We enter 100 in cell C5 (the total market size).

Step 2 We assume that originally Coke has half the market (see cell C6).

	A	B	C	D	E	F	G	H	I	J	K
1	**Brand share model**				Average						
2					Coke Share						
3	Coke switchers		0.1		66.1923077						
4	Pepsi switchers		0.2								
5	Market Size		100				Simulation Results for Book1				
6	Initial Coke share		0.5								
7							Iterations= 100				
8	Week	Coke share	Pepsi share	Leave Coke	Leave Pepsi		Simulations= 1				
9	1	50	50	6	8		# Input Variables= 104				
10	2	52	48	5	11		# Output Variables= 1				
11	3	58	42	6	11		Sampling Type= Latin Hypercube				
12	4	63	37	8	7		Runtime= 00:00:59				
13	5	62	38	7	9						
14	6	64	36	6	5		Summary Statistics				
15	7	63	37	3	6						
16	8	66	34	3	7		Cell	Name	Minimum	Mean	Maximum
17	9	70	30	5	7						
18	10	72	28	4	9		E3	Coke loyalty/C	61.8269	65.5929	68.5192
19	11	77	23	6	4						
20	12	75	25	6	4						
21	13	73	27	8	5						
22	14	70	30	9	4						
23	15	65	35	5	1						
24	16	61	39	9	8						
25	17	60	40	5	10						
26	18	65	35	2	13						
27	19	76	24	10	4						
28	20	70	30	9	6						
29	21	67	33	4	7						
30	22	70	30	8	6						
31	23	68	32	10	7						
32	24	65	35	6	8						
33	25	67	33	7	7						
34	26	67	33	6	6						
35	27	67	33	5	5						
36	28	67	33	10	5						
37	29	62	38	8	8						
38	30	62	38	7	8						
39	31	63	37	3	6						
40	32	66	34	7	5						
41	33	64	36	6	2						
42	34	60	40	6	8						
43	35	62	38	5	6						
44	36	63	37	5	8						
45	37	66	34	4	7						
46	38	69	31	7	10						
47	39	72	28	4	9						
48	40	77	23	7	4						
49	41	74	26	8	4						
50	42	70	30	7	7						
51	43	70	30	10	6						
52	44	66	34	10	5						
53	45	61	39	5	3						
54	46	59	41	5	6						
55	47	60	40	3	8						
56	48	65	35	4	7						
57	49	68	32	6	4						
58	50	66	34	2	11						
59	51	75	25	8	5						
60	52	72	28	5	4						

Figure 20.1 Basic Brand Share Model

Step 3 The fraction of Coke and Pepsi customers who switch to the other soda each week is given in cells C3 and C4.

Step 4 In cell B9 we compute Coke's Week 1 market share with the formula

```
Excel: =C5*C6

Lotus: +C5*C6
```

Step 5 In cell C9 we compute Pepsi's Week 1 market share with the formula

```
Excel: =C5-B9

Lotus: +C5-B9
```

Step 6 In cell D9 we model the number of Coke customers who will switch to Pepsi during Week 1 with the formula

```
Excel: =IF(B9=0,0,RISKBINOMIAL(B9,C$3)

Lotus: @IF(B9=0,0,RISKBINOMIAL(B9,C$3)
```

This indicates that each of the Coke purchasers (the current number of Coke purchasers is in cell B9) has a .1 chance of switching to Pepsi next week. The `IF` statement is needed because the `=RISKBINOMIAL` function is not defined when its first argument is less than or equal to 0.

Step 7 Similarly, in cell E9 we enter the formula

```
Excel: =IF(C9=0,0,RISKBINOMIAL(C9,C$4)

Lotus: @IF(C9=0,0,RISKBINOMIAL(C9,C$4)
```

to model the number of Pepsi customers who switch to Coke.

Step 8 In cell B10 we determine the number of Coke purchasers during week 2 with the formula

```
Excel: =B9-D9+E9

Lotus: +B9-D9+E9
```

This formula reflects the fact that

(Week 2 Coke Purchasers) = (Week 1 Coke Purchasers)
– (Coke Purchasers Who Switch to Pepsi)
+ (Pepsi Purchasers Who Switch to Coke)

Step 9 In cell C10 we determine the number of Week 2 Pepsi purchasers with the formula

```
Excel: =$C$5-B10

Lotus: $C$5-B10
```

This reflects that each customer buys each week from either Coke or Pepsi.

We now copy from the range B10:C10 to the range B10:C60 and from the range D9:E9 to the range D9:E60. This generates 52 weeks of sales.

Step 10 In cell E3 we compute average weekly sales for Coke with the formula

```
Excel:  =AVERAGE(B9:B60)

Lotus:  @AVG(B9..B60)
```

Step 11 Simulate this spreadsheet 100 times (with output cell E3).

We find that Coke sells an average of 65.592 cases per week and Pepsi sells an average of 34.408 cases per week.

Remark

It can be shown that after a large number of weeks, there is a 2/3 chance that a given customer will purchase Coke and a 1/3 chance that a customer will purchase Pepsi. We obtained a market share of < 2/3 for Coke because we assumed that initially Coke had < 2/3 of the market.

20.1 Is Advertising Worthwhile?

We now show how the modeling methodology developed in Example 20.1 can be used to analyze the effectiveness of advertising.

Example 20.1b

Suppose that the entire cola market consists of 100 million customers. A profit of $1 per case is earned. For $500 million per year, an advertising firm guarantees that they can decrease to 5% the percentage of Coke customers switching to Pepsi each week. Is this worthwhile?

Solution Figure 20.2 (file Cola2.wk4 or Cola2.xls) contains the relevant analysis. To obtain the market share for Coke after advertising we modify the entry in cell C3 to .05. Running 100 iterations of this spreadsheet we find Coke's average market share to be 77.565%. We can now compute the expected annual profit for Coke, with and without advertising.

Without Advertising Annual Expected Profit

$$= 52 * 100{,}000{,}000 * (.65592) = \$3.4107 \text{ billion}$$

With Advertising Annual Expected Profit

$$= 52 * 100{,}000{,}000 * (.77565) - 500{,}000{,}000 = \$3.5334 \text{ billion}$$

Thus we see that the advertising campaign is worthwhile!

	A	B	C	D	E	F	G	H	I	J	K
1	*				Average						
2	**with advertising**				Coke Share						
3	Coke switchers		0.05		76.88461538						
4	Pepsi switchers		0.2								
5	Market Size		100								
6	Initial Coke share		0.5								
7											
8	Week	Coke share	Pepsi share	Leave Coke	Leave Pepsi						
9	1	50	50	3	12		Simulation Results for Book1				
10	2	59	41	2	9						
11	3	66	34	7	8		Iterations= 100				
12	4	67	33	3	5		Simulations= 1				
13	5	69	31	3	7		# Input Variables= 104				
14	6	73	27	3	4		# Output Variables= 2				
15	7	74	26	2	7		Sampling Type= Latin Hypercube				
16	8	79	21	4	2		Runtime= 00:01:00				
17	9	77	23	7	3						
18	10	73	27	3	5		Summary Statistics				
19	11	75	25	2	4						
20	12	77	23	4	4		Cell	Name	Minimum	Mean	Maximum
21	13	77	23	2	6						
22	14	81	19	4	3						
23	15	80	20	4	1		!E3	Coke switchers/Coke .	74.26923	77.565	80.40385
24	16	77	23	1	6						
51	43	75	25	4	4						
52	44	75	25	7	7						
53	45	75	25	5	7						
54	46	77	23	6	2						
55	47	73	27	3	8						
56	48	78	22	2	3						
57	49	79	21	10	3						
58	50	72	28	0	4						
59	51	76	24	4	6						
60	52	78	22	4	4						

Figure 20.2 Brand Share Model with Advertising

20.2 To Coupon or Not to Coupon?

We now investigate whether Coke should give out coupons. A coupon sacrifices current profits to increase market share in the near future.

Example 20.1c

Coke is considering issuing a coupon for a case of soda every five weeks. The coupon is for a $.25 discount. Coke estimates that during the week the coupon is issued, only 5% of all Coke drinkers will switch to Pepsi and 40% of all Pepsi drinkers will switch to Coke. Will this couponing strategy increase profits? Assume that the cost of issuing coupons is negligible.

Solution The relevant spreadsheet is shown in Figure 20.3 (worksheet Cola3.wk4 or Cola3.xls). We proceed as follows:

Step 1 Enter the relevant switching probabilities in the cell range C3:D4 and the per case profit in the cell range C7:D7.

We assume that coupons are issued in Weeks 1, 6, 11, . . . , 54, 59.

Step 2 To model the couponing we enter into cell F9 the formula

```
Excel: =IF(MOD(A9,5)=1,1,0)

Lotus: @IF(@MOD(A9,5)=1,1,0)
```

The `@MOD` function returns the remainder when A9 is divided by 5. Thus our formula will enter a 1 in the rows corresponding to the weeks in which a coupon is issued, and a 0 in all other weeks. Then we copy this formula to the range F9:F60.

Step 3 To model the number of Coke drinkers who switch to Pepsi we enter in cell D9 the formula

```
Excel: =IF(B9=0,0,IF(F9=0,RISKBINOMIAL(B9,C$3),
RISKBINOMIAL(B9,D$3)))

Lotus: @IF(B9=0,0,@IF(F9=0,@RISKBINOMIAL
(B9,C$3),@RISKBINOMIAL(B9,D$3)))
```

This ensures that during a week in which a coupon is issued an average of 5% of Coke drinkers switch to Pepsi and that during a week in which a coupon is not issued an average of 10% of all Coke drinkers switch to Pepsi. We now copy this formula to the range D9:D60.

	A	B	C	D	E	F	G	H	I	J	K
1	**Brand share model**				Average						
2	**with coupons**		No coupon	Coupon	Weekly Profit						
3	Coke switchers		0.1	0.05	68.74038462						
4	Pepsi switchers		0.2	0.4							
5	Market Size		100								
6	Initial Coke share		0.5								
7	Profit per case		$ 1.00	$ 0.75							
8	Week	Coke share	Pepsi share	Leave Coke	Leave Pepsi	Coupon?	Profit	Simulation Results for COLA3.XLS			
9	1	50	50	2	19	1	37.5				
10	2	67	33	4	5	0	67	Iterations= 100			
11	3	68	32	3	4	0	68	Simulations= 1			
12	4	69	31	4	8	0	69	# Input Variables= 209			
13	5	73	27	7	6	0	73	# Output Variables= 1			
14	6	72	28	3	9	1	54	Sampling Type= Latin Hypercube			
15	7	78	22	6	2	0	78	Runtime= 00:00:54			
16	8	74	26	7	8	0	74				
17	9	75	25	6	7	0	75	Summary Statistics			
18	10	76	24	9	7	0	76				
19	11	74	26	3	7	1	55.5	Cell	Name	Minimum	Mean
20	12	78	22	7	7	0	78				
21	13	78	22	6	5	0	78	!E3	Weekly Profit	65.00961	68.47015
22	14	77	23	9	3	0	77				
23	15	71	29	4	4	0	71				
24	16	71	29	3	12	1	53.25				
51	43	60	40	8	9	0	60				
52	44	61	39	7	10	0	61				
53	45	64	36	5	6	0	64				
54	46	65	35	6	11	1	48.75				
55	47	70	30	0	5	0	70				
56	48	75	25	10	5	0	75				
57	49	70	30	7	3	0	70				
58	50	66	34	5	7	0	66				
59	51	68	32	2	14	1	51				
60	52	80	20	7	3	0	80				

Figure 20.3 Brand Share Model with Coupons

Step 4 In a similar fashion we model the number of Pepsi drinkers who switch to Coke by entering in cell E9 the formula

```
Excel: =IF(C9=0,0,IF(F9=0,RISKBINOMIAL(C9,C$4),
RISKBINOMIAL(C9,D$4)))

Lotus: @IF(C9=0,0,@IF(F9=0,@RISKBINOMIAL(C9,C$4),
@RISKBINOMIAL(C9,D$4)))
```

Copying this formula to the range E9:E60 generates the number of Pepsi drinkers switching to Coke each week.

All that remains is to compute Coke's weekly profit.

Step 5 We enter into cell G9 the formula

```
Excel: =IF((F9=0),C$7*B9,D$7*B9)

Lotus: @IF(F9=0,C$7*B9,D$7*B9)
```

Copying this formula to the range G9:G60 generates Coke's profit for each week.

We are assuming Coke earns a $.75 ($1.00 – $0.25) profit on each customer who purchases Coke during the week a coupon is issued. During all other weeks Coke earns a $1.00 profit per purchase.

Step 6 In cell E3 we compute average weekly profit for the year with the formula

```
Excel: =AVERAGE(G9:G60)

Lotus: @AVG(G9..G60)
```

Step 7 Select cell E3 as an output cell and use @RISK to simulate this spreadsheet 100 times.

We found that average weekly profit was $68.47. Thus the coupons appeared to increase average weekly profit (over the no advertising case).

Problems

Group A

20.1 Evaluate the option of Coke issuing coupons every three weeks.

20.2 Freezco sells refrigerators. Any refrigerator that fails before it is three years old must be replaced. Of all refrigerators 3% fail during their first year of operation; 5% of all one-year-old refrigerators fail during their second year of operation; and 7% of all two-year-old refrigerators fail during their third year of operation.

a Estimate the fraction of all refrigerators that will have to be replaced.

b It costs \$500 to replace a refrigerator, and Freezco sells 10,000 refrigerators per year. If the warranty period were reduced to two years, how much per year in replacement costs would be saved?

Group B

20.3 Seas Beginning sells clothing by mail order. An important question is when to strike a customer from their mailing list. At present, they strike a customer from their mailing list if a customer fails to order from six consecutive catalogs. They want to know if striking a customer from their list after a customer fails to order from four consecutive catalogs will result in a higher profit per customer. The following data is available:

If a customer placed an order the last time he received a catalog, then there is a 20% chance he will order from the next catolog.

If a customer last placed an order one catalog ago, there is a 16% chance she will order from the next catalog she receives.

If a customer last placed an order two catalogs ago, there is a 12% chance he will place an order from the next catalog he receives.

If a customer last placed an order three catalogs ago, there is an 8% chance she will place an order from the next catalog she receives.

If a customer last placed an order four catalogs ago, there is a 4% chance he will place an order from the next catalog he receives.

If a customer last placed an order five catalogs ago, there is a 2% chance she will place an order from the next catalog she receives.

It costs \$1 to send a catalog, and the average profit per order is \$15. Assume a customer has just placed an order. In order to maximize expected profit per customer, would Seas Beginning make more money canceling such a customer after six non-orders or four non-orders?

20.4 (Based on Babich 1992) Suppose that each week every family in the United States buys a gallon of orange juice from Company A, B, or C. Let p_I = probability that a gallon produced by company I is of unsatisfactory quality. If the last gallon of juice purchased by a family is satisfactory, then the next week they will purchase a gallon of juice from the same company. If the last gallon of juice purchased by a family is not satisfactory, then the family will purchase a gallon from a competitor. Consider a week in which A families have purchased Juice A, B families have purchased Juice B, and C families have purchased Juice C. Assume that families that switch brands during a period are allocated to the remaining brands in a manner that is proportional to the current market shares of the other brands. Thus if a customer switches from Brand A there is a chance $B/(B + C)$ that they will switch to Brand B and a chance $C/(B + C)$ that they will switch to Brand C. Suppose that 1,000,000 gallons of orange juice are purchased each week.

a After a long time, what will be the market share for each firm? Hint: You will need to use the `@RISKBINOMIAL` function to see how many people switch from A and then use the `@RISKBINOMIAL` function again to see how many switch from A to B and from A to C.

b Suppose a 1% increase in market share is worth $10,000 per week to Firm A. Also suppose that currently $p_A = .10$, $p_B = .15$, and $p_C = .20$. Firm A believes that for a cost of $1 million per year they can cut the percentage of unsatisfactory juice cartons in half. Is this worthwhile?

20.5 (Based on Hoppensteadt and Peskin 1992) The following model (the Reed-Frost model) is often used to model the spread of an infectious disease. Suppose that at the beginning of Period 1 the population consists of 5 diseased people (called infectives) and 95 healthy people (called susceptibles). During any period, there is a .05 chance that a given infective person will encounter a particular susceptible. If an infective encounters a susceptible, there is a .5 chance that the susceptible will contract the disease. An infective lives an average of 10 periods with the disease. To model this, we assume that there is a .10 probability that an infective dies during a period. Use @RISK to model the evolution of the population over 100 periods. Use your results to answer the following questions:

a What is the probability that the population will die out?

b What is the probability that the disease will die out?

c On the average, what percentage of the population becomes infected by the end of period 100?

d Suppose that people use infection "protection" during encounters. The use of protection reduces from .50 to .10 the chance that a susceptible will contract the disease during a single encounter with an infective. Now answer questions (a)–(c) under the assumption that everyone uses protection.

Hint: During any period there is a probability (.05)*.50 = .025 that an infective will infect a particular susceptible. Thus the probability that a particular susceptible is not infected during a period is $(1 - .025)^I$, where I = number of infectives present at end of the previous period.

20.6 Two brands of brownies are currently on the market. The sweetness, chewiness, and price of each brand (on a scale of 1 to 10) is given in Table 20.1.

Table 20.1

	Sweetness	Chewiness	Price
Brand 1	8	6	$3.00
Brand 2	10	7	$3.80

You are considering entering the brownie market. When buying brownies, an individual buys the brand with the highest score. The score of a brand is given by

(Individual's Sweetness Weight) * (Sweetness of Brand)
+ (Individual's Chewiness Weight) * (Chewiness of Brand)
+ (Individual's Price Weight) * (Price of brand in dollars).

We know that over all possible customers, the weights of individuals are normally distributed with the mean and standard deviation given in Table 20.2.

Table 20.2

	Mean	Standard Deviation
Sweetness Weight	5	1.0
Chewiness Weight	4	.6
Price Weight	–9	2.0

You are considering marketing a new brand of brownies. Brownies with the three specifications listed in Table 20.3 are under consideration.

Table 20.3

	Sweetness	Chewiness	Price
New Brand 1	8	6	$2.00
New Brand 2	10	9	$4.80

Which new brand would maximize your market share? Give a 95% confidence interval for the market share of the brand you choose to introduce.

20.7 You are unemployed and 21 years old and searching for a job. Until you accept a job offer the following situation occurs. At the beginning of each year you receive a job offer. The annual salary associated with the job offer is equally likely to be any number between $0 and $100,000. You must immediately choose whether to accept the job offer. If you accept an offer with salary $$x$, you receive $$x$ per year for the rest of your life (we assume you die at age 70) including the current year. Assume that cash flows are discounted so that a cash flow received one year from now has an NPV of .9. You have adapted the following policy: you will accept the first job offer that exceeds w^* dollars. Within $10,000, what value of w^* maximizes the expected amount of money you will receive the rest of your life?

References

Babich, P. 1992. "Customer Satisfaction: How Good is Good Enough?" *Quality Progress* 25 (Dec.): 65–68.

Hoppensteadt, F., and C. Peskin. 1992. *Mathematics in Medicine and the Life Sciences.* New York: Springer-Verlag.

CHAPTER 21

Generating Correlated Variables: Designing New Product

In Chapter 19 we used the RISKINDEPC and RISKDEPC commands to generate a pair of random variables with a given correlation. In this section we learn how to use the RISKCORRMAT command to generate a group of random variables with a given correlation matrix. To illustrate the idea, we consider the problem of determining the attributes of a new product that will maximize the product's market share. Our model is based on the work of Hauser and Gaskin (1984).

Example 21.1

Two brands of brownies are currently on the market. The sweetness, chewiness, and price of each brand (on a scale of 1 to 10) is given in Table 21.1.

Table 21.1

	Sweetness	Chewiness	Price
Brand 1	8	6	$3.00
Brand 2	10	7	$3.80

You are considering entering the brownie market. When buying brownies, an individual buys the brand with the highest score. An individual scores a brand as follows:

(Individual's Sweetness Weight) * (Sweetness of Brand)
+ (Individual's Chewiness Weight) * (Chewiness of Brand)
+ (Individual's Price Weight) * (Price of brand in dollars)

From a survey of potential brownie purchasers we estimate that the weights of individuals are normally distributed with the mean and standard deviation given in Table 21.2.

Table 21.2

	Mean	Standard Deviation
Sweetness Weight	5	1.0
Chewiness Weight	4	.6
Price Weight	–9	2.0

A survey of brownie purchasers indicates that the three weights are correlated as follows:

	Sweetness	Chewiness	Price
Sweetness	1.0	.8	.7
Chewiness	.8	1.0	.65
Price	.7	.65	1.0

You are considering marketing a new brand of brownies. Brownies with the three specifications listed in Table 21.3 are under consideration. Which new brand would maximize your market share?

Table 21.3

	Sweetness	Chewiness	Price
New Brand 1	8	6	$2.00
New Brand 2	10	9	$4.80

Solution Our simulation is illustrated in Figure 21.1 (see file Brownie1.wk4 or Brownie1.xls). We proceed as follows:

Step 1 In the cell range C6:D8 we entered the mean and standard deviation for the three weights (see Figure 21.1a).

Step 2 In the cell range C12:E14 we entered the correlation matrix for the weights.

Step 3 In the cell range C17:E20 we enter the "specifications" for the two brownie brands already in the market and for our two possible brands.

Step 4 In the cell range C23:E23 we generate the sweetness, chewiness, and price weights for an individual, based on the the given means, standard deviations, and correlation matrix. In cell C23 we generate the sweetness weight for an individual with the formula

	A	B	C	D	E	F	G	H	I
1	**New Product**		**Figure 21.1**						
2	**Decision with**								
3	**Correlated Preferences**								
4									
5	**Data for weights**		Mean Wt.	Wt. St. Dev.					
6		Sweetness	5	1					
7		Chewiness	4	0.6					
8		Price	-9	2					
9	**Correlation Matrix**								
10	**for Weights**								
11			Sweetness	Chewiness	Price				
12		Sweetness	1	0.8	0.7				
13		Chewiness	0.8	1	0.65				
14		Price	0.7	0.65	1				
15	**Product Specs**								
16			Sweetness	Chewiness	Price				
17		Brand 1	8	6	$ 3.00				
18		Brand 2	10	7	$ 3.80				
19		Our Choice 1	8	6	$ 2.00				
20		Our Choice 2	10	9	$ 4.80				
21									
22	**Generated Weights**		Sweetness	Chewiness	Price				
23			3.910162837	4.250503129	-9.089782				
24									
25		Scores							
26		Brand 1	29.51497696		Simulation Results for Book1				
27		Brand 2	34.31398056						
28		Our Choice 1	38.60475846		Iterations= 100				
29		Our Choice 2	33.72520531		Simulations= 1				
30					# Input Variables= 4				
31		Choice 1 bought?	1		# Output Variables= 3				
32		Choice 2 bought?	0		Sampling Type= Latin Hypercube				
33					Runtime= 00:00:18				
34									
35					Summary Statistics				
36									
37					Cell	Name	Minimum	Mean	Maximum
38									
39									
40					!C31	Choice 1 bought?!Sw...	0	0.68	1
41					!C32	Choice 1 bought?!Sw...	0	0.35	1

Figure 21.1 New Product Decision with Correlated Preferences

```
Excel: =RISKCORRMAT(C12:E14,1)+RISKNORMAL(C6,D6)

Lotus: +@RISKCORRMAT(C12..E14,1)+RISKNORMAL(C6,D6)
```

This formula ensures that the sweetness weight for an individual will follow a normal distribution with the desired mean and standard deviation. Also, the sweetness weight will be correlated with the other weights according to the weights in column 1 of the cell range C12:E14.

	A	B	C	D	E
1	**New Product**		**Figure 21.1**		
2	**Decision with**				
3	**Correlated Preferences**				
4					
5	**Data for weights**		Mean Wt.	Wt. St. Dev.	
6		Sweetness	5	1	
7		Chewiness	4	0.6	
8		Price	-9	2	
9	**Correlation Matrix**				
10	**for Weights**				
11			Sweetness	Chewiness	Price
12		Sweetness	1	0.8	0.7
13		Chewiness	0.8	1	0.65
14		Price	0.7	0.65	1
15	**Product Specs**				
16			Sweetness	Chewiness	Price
17		Brand 1	8	6	$ 3.00
18		Brand 2	10	7	$ 3.80
19		Our Choice 1	8	6	$ 2.00
20		Our Choice 2	10	9	$ 4.80
21					
22	**Generated Weights**		Sweetness	Chewiness	Price
23			3.9101628	4.250503129	-9.08978
24					
25		Scores			
26		Brand 1	29.514977		
27		Brand 2	34.313981		
28		Our Choice 1	38.604758		
29		Our Choice 2	33.725205		
30					
31		Choice 1 bought?	1		
32		Choice 2 bought?	0		

Figure 21.1a

Step 5 In cell D23 we generate the chewiness weight with the formula

```
Excel: =RISKCORRMAT(C12:E14,2)+RISKNORMAL(C7,D7)

Lotus: +@RISKCORRMAT(C12..E14,2)
+RISKNORMAL(C7,D7)
```

This formula ensures that the chewiness weight for an individual will follow a normal distribution with the desired mean and standard deviation. Also, the chewiness weight will be correlated with the other weights according to the weights in column 2 of the cell range C12:E14.

Step 6 In cell E23 we generate the price weight with the formula

```
Excel: =RISKCORRMAT(C12:E14,3)+RISKNORMAL(C8,D8)

Lotus: +@RISKCORRMAT(C12..E14,3)
+RISKNORMAL(C8,D8)
```

This formula ensures that the price weight for an individual will follow a normal distribution with the desired mean and standard deviation. Also, the price weight will be correlated with the other weights according to the weights in column 3 of the cell range C12:E14.

Step 7 In the cell range C26:C29 we now compute the individual's score for each type of brownie. In cell C26 we compute the individual's score for the competition's Brand 1 with the formula

```
Excel: =SUMPRODUCT(C$23:E$23,C17:E17)

Lotus: @SUMPRODUCT(C$23..E$23,C17..E17)
```

Copying this formula to the cell range C27:C29 computes the individual's score for the competition's Brand 2 and our two potential brands.

Step 8 Suppose we produce Brand 1. In cell C31 we determine whether the individual will purchase our Brand 1 with the formula

```
Excel: =IF(C28>MAX(C26:C27),1,0)

Lotus: @IF(C28>@MAX(C26..C27),1,0)
```

This formula will generate a 1 if an individual scores our Brand 1 higher than either of the competitors (to be conservative, we assume that in the case of a tie our product is not purchased).

Step 9 Similarly, suppose we produce Brand 2. In cell C32 we determine whether the individual will purchase our Brand 2 with the formula

```
Excel: =IF(C29>MAX(C26:C27),1,0)

Lotus: @IF(C29>@MAX(C26..C27),1,0)
```

Simulation Results for Book1				
Iterations= 100				
Simulations= 1				
# Input Variables= 4				
# Output Variables= 3				
Sampling Type= Latin Hypercube				
Runtime= 00:00:18				
Summary Statistics				
Cell	Name	Minimum	Mean	Maximum
!C31	Choice 1 bought?/Sw...	0	0.68	1
!C32	Choice 2 bought?/Sw...	0	0.35	1

Figure 21.1b

Step 10 Select the cell range C31:C32 as our output range and run 100 iterations of the spreadsheet with @RISK.

From the results (see Figure 21.1b) pasted into Figure 21.1, we find that Choice 1 would obtain a 68% market share, while Choice 2 would obtain only a 35% market share. Thus we should go with the lower quality, lower priced brownie, rather than the higher quality, higher priced brownie.

Remarks

1 The correlation matrix has a large effect on the market share for our Brand 2. To see this we used the following correlation matrix (see Figure 21.2, file Brownie2.wk4 or file Brownie2.xls):

	Sweetness	Chewiness	Price
Sweetness	1.0	–.80	.70
Chewiness	–.8	1.00	–.65
Price	.7	–.65	1.00

For this correlation matrix, we found that Brand 1 obtained a 67% market share, but Brand 2's market share dropped to 22% (see Figures 21.2a and 21.2b)!

	A	B	C	D	E	F	G	H	I
1	**New Product**		Revised	**Figure 21.2**					
2	**Decision with**		Correlation						
3	**Correlated Preferences**		Matrix						
4									
5	**Data for weights**		Mean Wt.	Wt. St. Dev.					
6		Sweetness	5	1					
7		Chewiness	4	0.6					
8		Price	-9	2					
9	**Correlation Matrix**								
10	**for Weights**								
11			Sweetness	Chewiness	Price				
12		Sweetness	1	-0.8	0.7				
13		Chewiness	-0.8	1	-0.65				
14		Price	0.7	-0.65	1				
15	**Product Specs**								
16			Sweetness	Chewiness	Price				
17		Brand 1	8	6	$ 3.00				
18		Brand 2	10	7	$ 3.80				
19		Our Choice 1	8	6	$ 2.00				
20		Our Choice 2	10	9	$ 4.80				
21									
22	**Generated Weights**		Sweetness	Chewiness	Price				
23			3.746801935	3.33372128	-9.236974				
24									
25		Scores							
26		Brand 1	22.26582185		Simulation Results for BROWNIE2.XLS				
27		Brand 2	25.70356799						
28		Our Choice 1	31.50279562		Iterations= 100				
29		Our Choice 2	23.13403678		Simulations= 1				
30					# Input Variables= 3				
31		Choice 1 bought?	1		# Output Variables= 2				
32		Choice 2 bought?	0		Sampling Type= Latin Hypercube				
33					Runtime= 00:00:10				
34									
35					Summary Statistics				
36									
37					Cell	Name	Minimum	Mean	Maximum
38									
39					C31	Choice 1 Bought	0	0.67	1
40					C32	Choice 2 Bought	0	0.22	1

Figure 21.2 New Product Decision with Correlated Preferences

Figure 21.2a

	A	B	C	D	E
1	**New Product**		Revised	**Figure 21.2**	
2	**Decision with**		Correlation		
3	**Correlated Preferences**		Matrix		
4					
5	**Data for weights**		Mean Wt.	Wt. St. Dev.	
6		Sweetness	5	1	
7		Chewiness	4	0.6	
8		Price	-9	2	
9	**Correlation Matrix**				
10	**for Weights**				
11			Sweetness	Chewiness	Price
12		Sweetness	1	-0.8	0.7
13		Chewiness	-0.8	1	-0.65
14		Price	0.7	-0.65	1
15	**Product Specs**				
16			Sweetness	Chewiness	Price
17		Brand 1	8	6	$ 3.00
18		Brand 2	10	7	$ 3.80
19		Our Choice 1	8	6	$ 2.00
20		Our Choice 2	10	9	$ 4.80
21					
22	**Generated Weights**		Sweetness	Chewiness	Price
23			3.7468019	3.33372128	-9.23697
24					
25		Scores			
26		Brand 1	22.265822		
27		Brand 2	25.703568		
28		Our Choice 1	31.502796		
29		Our Choice 2	23.134037		
30					
31		Choice 1 bought?	1		
32		Choice 2 bought?	0		

2 Most investment portfolios consist of holdings of several highly correlated investments. @RISK can be used to estimate the probability distribution of the return on such a portfolio. See Problem 21.1.

3 @RISK can also be used to value options involving correlated securities. Just assume that all securities underlying the option have an average growth rate equal to the risk-free rate. See Problem 21.2.

Figure 21.2b

Simulation Results for BROWNIE2.XLS				
Iterations= 100				
Simulations= 1				
# Input Variables= 3				
# Output Variables= 2				
Sampling Type= Latin Hypercube				
Runtime= 00:00:10				
Summary Statistics				
Cell	Name	Minimum	Mean	Maximum
C31	Choice 1 Bought	0	0.67	1
C32	Choice 2 Bought	0	0.22	1

4 You can also create correlated random variables by clicking on the Correlate button on the Input by Output List. You will then be prompted for the entries in the correlation matrix.

Problems

Group A

21.1 Suppose you have invested 25% of your portfolio in four different stocks. The mean and standard deviation of the annual return on each stock is as follows:

Table 21.4

	Mean	Standard Deviation
Stock 1	15%	20%
Stock 2	10%	12%
Stock 3	25%	40%
Stock 4	16%	20%

The correlation between the annual returns on the four stocks is as follows:

	Stock 1	Stock 2	Stock 3	Stock 4
Stock 1	1.00	.80	.70	.60
Stock 2	.80	1.00	.75	.55
Stock 3	.70	.75	1.00	.65
Stock 4	.60	.55	.65	1.00

a What is the probability that my portfolio's annual return will exceed 20%?

b What is the probability that my portfolio will lose money during the course of a year?

21.2 Suppose that the current price of each stock in Problem 21.1 is as follows:

Stock 1: $14 Stock 3: $18

Stock 2: $16 Stock 4: $20

I have just bought an option involving these four stocks. If the price of Stock 1 six months from now is $17 or more, the option enables me to buy, if I desire, one share of each stock for $20 six months from now. Otherwise the option is worthless. For example, if six months from now the stock prices are

Stock 1: $18 Stock 3: $21

Stock 2: $19 Stock 4: $22

then I would exercise my option to buy Stocks 3 and 4 and receive (21 – 20) + (22 – 20) = $3 in cash flow. How much is this option worth if the risk-free rate is 8%?

21.3 There are three market segments in the computer industry. Each consists of 100,000 customers. Customers choose between computers based on three attributes: Power/Speed, Design, and Price. The mean and standard deviation of the weights customers in each segment give to each attribute are normally distributed with the following means and standard deviations:

Table 21.5

Attribute	Customer Segment	Mean	Standard Deviation
Power/Speed	1	9	2.0
Power/Speed	2	5	4.5
Power/Speed	3	4	1.0
Design	1	2	.5
Design	2	7	3.0
Design	3	5	2.0
Price	1	–4	4.0
Price	2	–6	1.5
Price	3	–8	.8

For each segment, the correlation between attribute weights is as follows:

	Power/Speed	Design	Price
Power/Speed	1.0	.2	.8
Design	.2	1.0	.4
Price	.8	.4	1.0

Suppose four computers with the following attributes are on the market (10 = powerful computer, excellent design, or high price):

Table 21.6

	Power/Speed	Design	Price
Computer 1	10	8	10
Computer 2	8	6	7
Computer 3	7	3	5
Computer 4	4	2	3

Estimate the market share for each computer.

Group B

21.4 You are thinking about purchasing an option that gives you the opportunity to exchange one year from now 1 oz. of gold for 100 oz. of silver. Currently gold sells for $380 per oz. and silver sells for $4 per oz. The volatility for the change in the price of gold and silver during a three-month period is 10%. The correlation between the change in gold and silver prices is .7. The risk-free rate is 10%. We estimate that the mean change in the price of gold during the next year is 15% and the mean change in the price of silver is 20%. What is a fair price to pay for this option? Assume that (16.1) can be used to model the growth of gold and silver prices.

21.5 The B-School at State U currently has three parking lots each containing 155 spaces. Two hundred faculty have been assigned to each lot. On a peak day, an average of 70% of all Lot 1 parking sticker holders show up, an average of 72% of all Lot 2 parking sticker holders show up, and an average of 74% of all Lot 3 parking sticker holders show up.

a Given the current situation, estimate the probability that on a peak day at least one faculty member with a sticker will be unable to find a spot. Assume that the number who show up at each lot is independent of the number who show up at the other two lots. Can you come up with a solution to this problem (that does not involve creating more parking spaces!)?

b Now suppose the number of people who show up at the three lots are correlated (correlation .9). Does your solution work as well? Why or why not?

21.6 Bottleco produces six-packs of soda cans. Each can is supposed to contain at least 12 oz. of soda. If the total weight in a six-pack is under 72 oz., Bottleco is fined $100 and receives no sales revenue for the six-pack. Each six-pack sells for $3.00. It costs Bottleco $.02 per ounce of soda put in the cans. Bottleco can control the mean fill rate of its soda-filling machines. The amount put in each can by a machine is normally distributed with a standard deviation of .10 oz.

a Assume that the weight of each can in a six-pack has a .8 correlation with the weight of the other cans in the six-pack. What mean fill quantity (within .05 oz.) maximizes expected profit per six-pack?

b If the weights of the cans in the six-pack are independent, what mean fill quantity (within .05 oz.) will maximize expected profit per six-pack?

c Can you explain the difference in the answers to parts (a) and (b)?

Reference

Hauser, J., and S. Gaskin. 1984. "Application of the Defender Consumer Model." *Marketing Science* (vol. 3 no. 4): 327–351.

CHAPTER 22

Simulating Sample Plans with the Hypergeometric Distribution

Often companies receive a shipment of material from a supplier and need to ascertain the quality of the shipment. If it is impractical to inspect every item in the shipment, then a sampling plan is often developed to help determine the quality of the shipment. A common sampling plan is the **(*n, c*) single sampling plan**. In an (*n, c*) plan, *n* items are chosen (without replacement) from the batch of shipped material. If *c* or fewer of the sampled items are defective, then the batch is accepted, otherwise the batch is rejected.

Sampling plans have two types of risk associated with them; producer's risk and consumer's risk. Before defining producer's and consumer's risk we need to define the **Acceptable Quality Level (AQL)** and **Lot Tolerance Percent Defective (LTPD)**. The AQL is the maximum percent defective that can be considered satisfactory as a process average. The LTPD is a percent defective for which the consumer would want a batch to be regularly rejected. The **producer's risk** for a sampling plan can now be defined as the probability of rejecting a batch whose quality actually equals AQL. That is, the producer's risk is the chance that a satisfactory batch (defined by AQL) is rejected by a sampling plan. The **consumer's risk** for a sampling plan is the probability that an unsatisfactory batch (as defined by LTPD) is accepted.

The number of defectives in a sample (assuming that sampling is conducted without replacement) follows the *hypergeometric distribution*. The @RISK function

```
Excel:  =RISKHYPERGEO(n,d,N)

Lotus:  @RISKHYPERGEO(n,d,N)
```

will generate the number of defectives in a sample of size *n* from a batch of *N* items, *d* of which are defective.

Example 22.1 shows how to use @RISK to analyze the effectiveness of a sampling plan.

Example 22.1

A company receives a batch of 10,000 computers. They are going to use an (89, 2) sampling plan. That is, a sample of 89 computers is inspected, and the entire batch of 10,000 computers is accepted if 0, 1, or 2 defective computers are found in the sample. Otherwise the batch is rejected. If AQL = 1% and LTPD = 5%, what are the consumer's and producer's risks for this sampling plan?

Solution We proceed as follows:

Step 1 In the cell range D4:D7 (see Figure 22.1 and file Sampling.xls or Sampling.wk4) we enter the parameters of the sampling plan and the batch size.

Step 2 In the cell range D8:D9 we enter the AQL and LTPD.

Step 3 In cell D10 we enter the fraction of the batch that we assume to be defective corresponding to AQL and LTPD with the formula

```
Excel:  =RISKSIMTABLE(D8:D9)

Lotus:  @RISKSIMTABLE(D8,D9)
```

Note that when in Excel @RISK, we refer to a cell range when using an array @RISK function such as `RISKSIMTABLE` or `RISKDISCRETE;` we must type the cell range as a range, not a sequence of cells. Also, we must leave out the brackets { } around the arguments for `RISKSIMTABLE.`

Note that the entire batch will contain D10*D4 defectives.

Step 4 In cell B12 we generate the number defective in the sample with the formula

```
Excel:  =RISKHYPERGEO(D5,D10*D4,D4)

Lotus:  @RISKHYPERGEO(D5,D10*D4,D4)
```

Step 5 In cell B14 we determine whether the sample results in an accepted batch with the formula

```
Excel:  =IF(B12<=D7,1,0)

Lotus:  @IF(B12<=D7,1,0)
```

This formula will cause any batch that includes a sample in which two or fewer defectives are found to be accepted.

	A	B	C	D	E	F	G	H	I
1	**Calculating Producer's and Consumer's**					**Figure 22.1**			
2	**Risk in Sampling**				Simulation Results for Book2				
3									
4	Batch Size			10000	Iterations= 100				
5	Sample Size			89	Simulations= 2				
6	Accept lot if number of defectives is <=c				# Input Variables= 2				
7			c	2	# Output Variables= 1				
8	Acceptable Quality level(AQL)			0.01	Sampling Type= Latin Hypercube				
9	Lot tolerance percent defective(LTPD)			0.05	Runtime= 00:00:09				
10	Percentage defective in batch			0.01					
11	Number of defectives in Sample				Summary Statistics				
12		3							
13	Sample Accepted?				Cell	Name	Minimum	Mean	Maximum
14		0							
15					B14	(Sim#1) Accepted	0	0.94	1
16					B14	(Sim#2) Accepted	0	0.18	1
17					D10	(Sim#1) (Input) Perc...	0.01	0.01	0.01
18					D10	(Sim#2) (Input) Perc...	0.05	0.05	0.05

Figure 22.1 Calculating Consumer's and Producer's Risk

Step 6 We now make cell B14 our output cell and run two simulations (100 iterations each). Note that in our first simulation each item sampled had a 1% chance of being defective and in our second simulation each sampled item had a 5% chance of being defective.

We find that 94% of all batches with 1% defectives are accepted by the plan. This makes the producer's risk 1 – .94 = .06, because 6% of "good" batches will be rejected by the sampling plan.

We find that 18% of all batches with 5% defectives are accepted by the sampling plan. Hence the consumer's risk is 18%, because 18% of all "bad" batches will be accepted by the sampling plan.

In Problem 22.3 we discuss the more complex "double sampling" plan.

Problems

Group A

22.1 Suppose that AQL = .1 and LTPD = .2. For a batch of size 1000 evaluate the producer's and consumer's risk for the following plans:

a (3, 0)

b (5, 1)

c (100, 4)

22.2 Suppose that in any accepted lot the defectives in the sample are replaced with good items and in any rejected lot *all* items are replaced with good items. For a given sampling plan, the **Average Outgoing Quality (AOQ)** is defined to be the fraction of all outgoing items that are defective. For each of the sampling plans in Problem 22.1, estimate AOQ.

Group B

22.3 Consider the following **double sampling plan** for a batch of size 1000. Draw a sample of size 100. Accept the sample if 3 or fewer defectives are found and reject the sample immediately if more than 9 defectives are found. If 4–9 defectives are found, take another sample of size 200. If the total number of defectives in the two samples is 9 or less, accept the lot; otherwise reject the lot. If AQL = .02 and LTPD = .05, evaluate the consumer's and producer's risk for this double sampling plan.

CHAPTER 23

Simulating Inventory Models

Consider a company that faces uncertain demand for a product. The company must make two key decisions:

Reorder Quantity: How much should be ordered each time an order is placed?

Reorder Point: How low should the company let its inventory level go before it decides to place an order?

We will consider the **periodic review** inventory system, that is, a system in which a firm reviews its inventory at periodic intervals (say the beginning of each week) and determines whether to place an order.

The company incurs the following types of costs:

Fixed Ordering Cost: Whenever an order is placed, a fixed cost K is incurred. This cost is independent of the size of the order.

Unit Purchase Cost: A cost p per unit ordered is incurred.

Unit Holding Cost: A cost h is incurred for each unit in inventory at the end of a given time period.

Unit Shortage Cost: A cost s is incurred each time period for each unit of demand that is unmet at the end of a time period.

The firm is able to **backlog demand**, that is, demand that is unmet may be met during a later period.

The following example shows how @RISK can be used to simulate a periodic review inventory system. For a discussion of analytic models relevant to this situation see Chapter 17 of Winston (1994).

Example 23.1

A company has a periodic review (weekly) inventory system for a product having $K = \$200$, $p = \$4.00$, $h = \$3.00$, and $s = \$10.00$. Weekly demand for the product follows a normal distribution with a mean of 30 and a standard deviation of 6. The company is currently placing an order whenever the inventory at the beginning of the week is less than 50 units. If the inventory level equals x when an order is placed, a quantity $150 - x$ is ordered. An order is equally likely to arrive 1, 2, or 3 weeks after it is placed. We assume that if an order is in transit, no other orders will be placed. Use simulation to estimate the average weekly cost associated with this inventory policy. Assume that initially 100 units of inventory are in stock.

Solution Our work is in the file Inven.xls or Inven.wk4 (see Figure 23.1). We have chosen to simulate the system for 40 weeks.
For each month we keep track of the following quantities:

Beginning Inventory: This is the inventory after an order (if any) is received, but before demand is met.

Quantity on Order: This equals 0 if nothing is on order. Otherwise it is the size of the order that we are waiting for.

Next Order Received: If no order is in the pipeline, we set this equal to a large number (9999). Otherwise, this is the number of the week during which the order in the pipeline arrives. Of course, the arrival date of the order is random.

Demand: This is the weekly demand for the product and is, of course, random.

Ending Inventory: This is our inventory after all possible demand has been met. A negative ending inventory means that a shortage has occurred.

Cost: This is the sum of fixed ordering, purchase, holding, and shortage costs incurred during the week. We assume that fixed and purchase costs are incurred when an order is placed, not when it arrives.

We begin by keeping track of what happens during Week 1 (see Figure 23.1a).

Step 1 In the cell range A1:B6 we enter the parameters describing the problem's cost structure and our ordering policy.

Step 2 In B9 we enter Week 1's beginning inventory (100).

Step 3 In B10 we compute the Quantity on Order with the formula

	A	B	C	D	E	F	G	H	AL	AM	AN	AO
1	Shortage cost	$ 10.00		**Inventory**								
2	Holding cost	$ 3.00		**Model**	**Figure 23.1**							
3	Fixed cost	$ 200.00		**with**								
4	Unit Purchasing cost	$ 4.00		**Random**								
5	Order up to	150		**Leadtime**								
6	Reorder point	50										
7												
8	Week	1	2	3	4	5	6	7	37	38	39	40
9	Beginning inventory	100	59.0549	31.7206	-8.217914999	-40.6266	52.1593	17.3128	125.447	90.138	55.5953	20.8381
10	Quantity on order	0	0	118.2794	118.2794014	118.279	0	132.687	0	0	0	129.162
11	Next Order received	9999	9999	6	6	6	9999	9	9999	9999	9999	41
12	Demand	40.94509	27.3343	39.93851	32.40871835	25.4935	34.8465	25.7268	35.3091	34.5426	34.7572	30.3959
13	Ending Inventory	59.05491	31.7206	-8.21791	-40.62663335	-66.1201	17.3128	-8.414	90.138	55.5953	20.8381	-9.55775
14	Cost	177.1647	95.1618	755.2968	406.2663335	661.201	51.9384	814.889	270.414	166.786	62.5144	812.225
15												
16	Average Cost	332.7041			Variable Type							
17					Name	Average Cost						
18					Description	Output						
19					Cell	B16						
20					Minimum =	313.449						
21					Maximum =	423.247						
22					Mean =	359.327						
23					Std Deviation =	21.5051						
24					Variance =	462.467						

Figure 23.1 Inventory Simulation

	A	B	C	D	E	F	G	H	AL	AM	AN	AO
1	Shortage cost	$ 10.00		Inventory								
2	Holding cost	$ 3.00		Model	Figure 23.1							
3	Fixed cost	$ 200.00		with								
4	Unit Purchasing cost	$ 4.00		Random								
5	Order up to	150		Leadtime								
6	Reorder point	50										
7												
8	Week	1	2	3	4	5	6	7	37	38	39	40
9	Beginning inventory	100	59.0549	31.7206	-8.217914999	-40.6266	52.1593	17.3128	125.447	90.138	55.5953	20.8381
10	Quantity on order	0	0	118.2794	118.2794014	118.279	0	132.687	0	0	0	129.162
11	Next Order received	9999	9999	6	6	6	9999	9	9999	9999	9999	41
12	Demand	40.94509	27.3343	39.93851	32.40871835	25.4935	34.8465	25.7268	35.3091	34.5426	34.7572	30.3959
13	Ending Inventory	59.05491	31.7206	-8.21791	-40.62663335	-66.1201	17.3128	-8.414	90.138	55.5953	20.8381	-9.55775
14	Cost	177.1647	95.1618	755.2968	406.2663335	661.201	51.9384	814.889	270.414	166.786	62.5144	812.225

Figure 23.1a

```
Excel: =IF(B9<$B$6,$B$5-B9,0)
```

```
Lotus: @IF(B9<$B$6,$B$5-B9,0)
```

This ensures that an order is placed if and only if beginning inventory is less than the reorder point and that any order is for the proper amount. Since our Week 1 beginning inventory is > 50, no order is placed.

Step 4 We now keep track of the week the next order is received. If no order is placed, we make the time the next order is placed equal a large number (say 9999). If an order is placed, we generate the duration of the order's leadtime with the RISKDISCRETE function, and record the arrival date of the order. Entering the following statement in cell B11 accomplishes these goals.

```
Excel: =IF(B10=0,9999,B8
+RISKDISCRETE({1,2,3},{1,1,1})
```

```
Lotus: @IF(B10=0,9999,B8
+@RISKDISCRETE(1,2,3,1,1,1)
```

Step 5 We now generate the demand for Week 1 in cell B12 with the formula

```
Excel: =RISKNORMAL(30,6)
```

```
Lotus: @RISKNORMAL(30,6)
```

Step 6 The firm's ending inventory for Week 1 will equal its beginning inventory less its Week 1 demand. This is computed in cell B13 with the formula

```
Excel: =B9-B12
```

```
Lotus: +B9-B12
```

Step 7 In cell B14 we compute Week 1's total cost (fixed cost, purchase cost, holding cost, and shortage cost) with the formula

```
Excel: =IF(B10>0,$B$3,0)+$B$4
*B10+$B$2*MAX(B13,0)+$B$1
*MAX(-B13,0)
```

```
Lotus: @IF(B10>0,$B$3,0)+$B$4
*B10+$B$2*MAX(B13,0)+$B$1
*MAX(-B13,0)
```

The first term in the sum is the Week 1 fixed cost, the second term the Week 1 purchase cost, the third term the Week 1 holding cost, and the fourth term the Week 1 shortage cost.

We now determine what happens during Week 2.

Step 8 Week 2's beginning inventory will equal Week 1's ending inventory + any order received during Week 2. Entering the following formula in C9 ensures that we adjust Week 1's ending inventory by the size of an order if an order arrives.

```
Excel: =B13+IF(B11=C8,B10,0)

Lotus: +B13+@IF(B11=C8,B10,0)
```

Step 9 If the quantity ordered during Week 1 is scheduled to arrive later than Week 2, the Quantity on Order for Week 2 will equal the Quantity on Order for Week 1. If this is not the case, and we begin Week 2 with our inventory below the reorder point, an order must be placed. Otherwise, an order has arrived this week or nothing is on order, and the Week 2 Quantity on Order should equal 0. Entering the following nested `IF` statement in cell C10 operationalizes these relationships.

```
Excel: =IF(AND((B11<45),(B11>C8)),B10,
IF(C9<$B$6,$B$5-C9,0))

Lotus: @IF(AND((B11<45),(B11>C8)),B10,
@IF(C9<$B$6,$B$5-C9,0))
```

Step 10 If nothing is on order, we make the week the next order arrives equal to 9999. If an order has just been placed this week (this is the case only if |C10 – B10|>0!), then we must update the arrival of the next order by generating a random leadtime. Otherwise, we are still waiting for an order, and the value of Next Order Received remains unchanged. These relationships are captured in cell C11 with the formula

```
Excel: =IF(C10=0,9999,IF(ABS(C10-B10)>0,C8
+RISKDISCRETE({1,2,3},{1,1,1}),B11))

Lotus: @IF(C10=0,9999,IF(ABS(C10-B10)>0,C8
+@RISKDISCRETE(1,2,3,1,1,1),B11))
```

Step 11 In cell C12 we generate Week 2's demand with the formula

```
Excel: =RISKNORMAL(30,6)

Lotus: @RISKNORMAL(30,6)
```

Figure 23.1b

Variable Type	
Name	Average Cost
Description	Output
Cell	B16
Minimum =	313.4493
Maximum =	423.2474
Mean =	359.3273
Std Deviation =	21.50505
Variance =	462.467

Step 12 In cell C13 we compute Week 2's ending inventory with the formula

```
Excel: =C9-C12.

Lotus: +C9-C12
```

Step 13 In cell C14 we compute Week 2's cost with the formula

```
Excel: =IF(AND((C10>0,(B10=0)),$B$3+$B$4*C10,0)
+$B$2*MAX(C13,0)+$B$1*MAX(-C13,0)

Lotus: @IF(AND((C10>0,(B10=0)),$B$3+$B$4*C10,0)
+$B$2*MAX(C13,0)+$B$1*MAX(-C13,0)
```

`=IF(AND((C10>0,(B10=0)),$B$3+$B$4*C10,0)` captures the Week 2 fixed and purchase costs while `$B$2*MAX(C13,0)+$B$1*MAX(-C13,0)` captures the Week 2 holding and shortage costs.

Step 14 Copying from the cell range C9:C14 to the cell range D9:AO14 simulates 40 weeks of this inventory system.

Step 15 In cell B16 we compute the average weekly cost with the formula

```
Excel: =AVERAGE(B14:AO14)

Lotus: @AVG(B14:AO14)
```

Step 16 We selected cell B16 as our output range and ran 100 iterations of this simulation.

We found the average weekly cost for this policy to be \$359.33 (see Figure 23.1b). Note that the standard deviation of the average profits for our 100 iterations was only \$21.50, so a 95% confidence interval for weekly costs would be quite narrow!

Remark

Using the `=RISKSIMTABLE` command we could have @RISK "try out" many inventory policies.

Problems

Group A

23.1 Suppose that we modify Example 23.1 so that when a shortage occurs, the sale is lost and an additional penalty of $8 (due to lost profit from a sale) is incurred for each lost sale. Determine an estimate for expected weekly cost.

23.2 Computco sells personal computers. The worst-case monthly demand for computers is 300: the best-case monthly demand is 1000; and the most-likely monthly demand is 800 computers. Each time an order is placed, a cost of $600 per order and $1500 per computer is incurred. Computers are sold for $2800, and if Computco does not have a computer in stock, the customer will buy a computer from a competitor. At the end of each month a holding cost of $10 per computer is incurred. Orders are placed at the end of each month and arrive at the beginning of the next month. Four ordering policies are under consideration:

Policy 1: Place an order for 900 computers whenever the end-of-month inventory is 50 or less.

Policy 2: Place an order for 600 computers whenever the end-of-month inventory is 200 or less.

Policy 3: Place an order for 1000 computers whenever end-of-month inventory is 400 or less.

Policy 4: Place an order for 1200 computers whenever end-of-month inventory is 500 or less.

Using 500 iterations, determine which ordering policy will maximize expected profit for two years. To get a more accurate idea of expected profit you can credit yourself with a salvage value of $1500 for each computer left at the end of the last month.

Assume 400 computers are in inventory at the beginning of the first month.

Hint: You can use the `@RISKSIMTABLE` command to analyze all four ordering policies in a single spreadsheet.

23.3 Lowland Appliance replenishes its stock of color TVs three times a year. Each order takes 1/9 of a year to arrive. Annual demand for color TVs follows a normal distribution with a mean of 990 and a standard deviation of 40. Assume that the cost of holding a TV in inventory for a year is $100. Assume that we begin with 500 TVs in inventory, the cost of a shortage is $150, and the cost of placing an order is $500.

a Suppose that whenever inventory is reviewed and our inventory level is x, we order $480 - x$ TVs. Estimate the average annual cost of such a policy. Such a policy is called an **order-up policy**.

b Estimate average annual cost for order-up policies when we order up to 200, 400, 600, and 800 TVs.

Group B

23.4 A highly perishable drug spoils after three days. A hospital estimates that they are equally likely to need between 1 and 9 units of the drug daily. Each time an order for the drug is placed, a fixed cost of $200 is incurred as well as a purchase cost of $50 per unit. Orders are placed at the end of each day and arrive at the beginning of the following day. It costs no money to hold the drug in inventory, but a cost of $100 is incurred each time the hospital needs a unit of the drug and does not have any available. The following three policies are under consideration:

Policy 1: If we end the day with fewer than 5 units, order enough to bring next week's beginning inventory up to 10 units.

Policy 2: If we end the day with fewer than 3 units order enough to bring next week's beginning inventory up to 7 units.

Policy 3: If we end the day with fewer than 8 units order enough to bring next week's beginning inventory up to 15 units.

Compare these policies with regard to expected daily costs, expected number of units short per day, and expected number of units spoiling each day. Assume that we begin Day 1 with 5 units of the drug on hand.

Hint: You will need to keep track of the age distribution of the units on hand at the beginning of each week. Assume that the hospital uses a FIFO (First in, First out) inventory policy. The trick is to get formulas that relate the age of each unit of the drug you have at the beginning of the day to the age of each unit you have at the end of the day.

CHAPTER 24

Simulating a Single-Server Queuing System

Probably the most common application of simulation has been the study of waiting line or queuing systems. Unfortunately, @RISK is not well-suited for such simulations, so we will limit our discussion of queuing simulations to a single chapter. Most queuing simulations are done with special purpose packages such as GPSS, SIMSCRIPT, SIMFACTORY, SIMAAN, or SLAM. See the end-of-chapter references for some books that explain in detail how to use special packages for queuing (often called "bottleneck") simulations.

In our example, we will assume that the time between arrivals and a customer's service time is governed by an *exponential* random variable. A histogram for a typical exponential distribution would be shaped like Figure 24.1. Notice that large values are possible, but very unlikely. The function

```
Excel: =RISKEXPON(MEAN)

Lotus: @RISKEXPON(MEAN)
```

generates observations from an exponential random variable with a mean equal to `MEAN`.

The following example illustrates the general methodology used in most queuing simulations.

Example 24.1

A small fast-food restaurant has a single drive-through window. An average of 10 customers per hour arrive, and the drive-through window can service an average of

Figure 24.1	
Exponential Distribution	
with mean 6	
x	f(x)
0.1	0.163912
0.2	0.161203
0.3	0.158538
0.4	0.155918
0.5	0.153341
0.6	0.150806
0.7	0.148314
0.8	0.145862
0.9	0.143451
1	0.14108
1.1	0.138748
1.2	0.136455
1.3	0.1342
1.4	0.131982
1.5	0.1298
1.6	0.127655
1.7	0.125545
1.8	0.12347
1.9	0.121429
2	0.119422
2.1	0.117448
2.2	0.115507
2.3	0.113598
2.4	0.11172

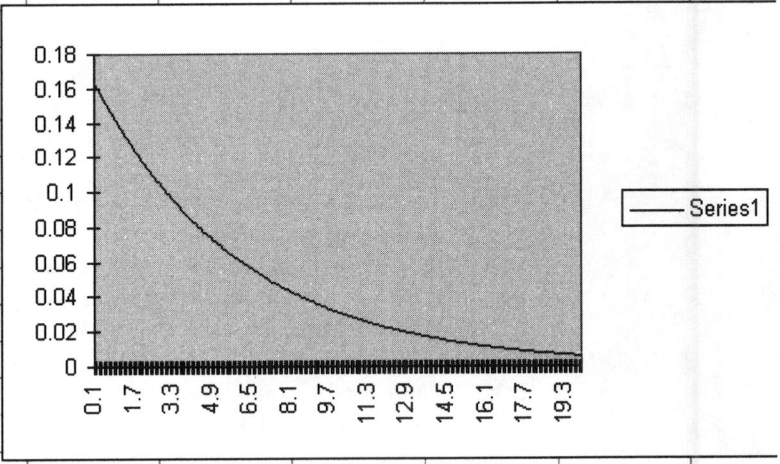

Figure 24.1 Exponential Distribution with Mean 6

	A	B	C	D	E	F	G	H	I	J	K
1	EVENT#	EVENT TYPE	INTER-ARRIVAL TIME	SERVICE TIME	TM	SS	WL	AT	DT	#IN SYS	TIME BETWEEN EVENTS
2	0				0	0		0	9999	0	
3	1	arrival	13.04836	9.309964	0	1	0	13.0484	9.30996	1	9.30996431
4	2	departure	6.503558	5.438063	9.30996	0	0	13.0484	9999	0	3.73839937
5	3	arrival	15.61003	7.235363	13.0484	1	0	28.6584	20.2837	1	7.23536307
6	4	departure	1.887665	7.685358	20.2837	0	0	28.6584	9999	0	8.37467043
7	5	arrival	7.387809	6.346526	28.6584	1	0	36.0462	35.0049	1	6.34652601
8	6	departure	1.993136	0.387931	35.0049	0	0	36.0462	9999	0	1.04128343
9	7	arrival	12.62617	7.793151	36.0462	1	0	48.6724	43.8394	1	7.79315123
10	8	departure	5.264808	3.026689	43.8394	0	0	48.6724	9999	0	4.83302061
11	9	arrival	17.67329	1.283547	48.6724	1	0	66.3457	49.9559	1	1.28354742
12	10	departure	12.4956	1.001414	49.9559	0	0	66.3457	9999	0	16.3897401
13	11	arrival	7.548844	4.764584	66.3457	1	0	73.8945	71.1102	1	4.76458391
14	12	departure	3.942093	1.419442	71.1102	0	0	73.8945	9999	0	2.78425995
15	13	arrival	1.027172	7.551005	73.8945	1	0	74.9217	81.4455	1	1.02717184
16	14	arrival	12.1762	0.277342	74.9217	1	1	87.0979	81.4455	2	6.52383281
17	15	departure	1.02616	1.627381	81.4455	1	0	87.0979	83.0729	1	1.62738077
18	16	departure	7.21177	2.712945	83.0729	0	0	87.0979	9999	0	4.02498403
19	17	arrival	8.405306	4.61078	87.0979	1	0	95.5032	91.7087	1	4.61078018
20	18	departure	0.405289	2.631934	91.7087	0	0	95.5032	9999	0	3.79452539
21	19	arrival	2.182983	3.317506	95.5032	1	0	97.6862	98.8207	1	2.18298329
22	20	arrival	6.979269	0.74126	97.6862	1	1	104.665	98.8207	2	1.13452306
23	21	departure	0.200413	1.557443	98.8207	1	0	104.665	100.378	1	1.55744323
24	22	departure	0.025372	3.055414	100.378	0	0	104.665	9999	0	4.2873024
25	23	arrival	4.247138	4.640461	104.665	1	0	108.913	109.306	1	4.24713781
26	24	arrival	21.12531	6.364922	108.913	1	1	130.038	109.306	2	0.39332302
27	25	departure	2.777447	4.049909	109.306	1	0	130.038	113.356	1	4.04990938
28	26	departure	4.843256	5.766867	113.356	0	0	130.038	9999	0	16.6820815
29	27	arrival	2.498226	2.987178	130.038	1	0	132.536	133.025	1	2.49822595
30	28	arrival	0.949637	0.111193	132.536	1	1	133.486	133.025	2	0.48895222
31	29	departure	5.493297	1.908002	133.025	1	0	133.486	134.933	1	0.46068473
32	30	arrival	10.32368	2.181354	133.486	1	1	143.809	134.933	2	1.44731687
33	31	departure	2.647277	1.798745	134.933	1	0	143.809	136.732	1	1.79874481
34	32	departure	0.965501	4.613521	136.732	0	0	143.809	9999	0	7.07761662
35	33	arrival	6.747961	2.561279	143.809	1	0	150.557	146.371	1	2.56127854
36	34	departure	1.639204	1.305297	146.371	0	0	150.557	9999	0	4.18668277
37	35	arrival	2.529057	1.724248	150.557	1	0	153.086	152.282	1	1.72424803
38	36	departure	13.67412	3.287086	152.282	0	0	153.086	9999	0	0.80480929
39	37	arrival	0.982895	9.371352	153.086	1	0	154.069	162.458	1	0.98289463
40	38	arrival	1.636461	4.405374	154.069	1	1	155.706	162.458	2	1.63646084
41	39	arrival	18.26191	1.757511	155.706	1	2	173.968	162.458	3	6.75199624
42	40	departure	18.67183	0.661129	162.458	1	1	173.968	163.119	2	0.66112929
43	41	departure	0.835073	1.144784	163.119	1	0	173.968	164.264	1	1.14478448

	L	M	N	O	P	Q
1	Fraction of time busy	Average Number in system				
2	0.66093	1.76661				
4	Queuing Simulation					
5		Figure 24.2				
8		Cell	Name	Minimum	Mean	Maximu
10		L2	/Fraction of ti	0.415183	0.6526	0.9686
11		M2	/Fraction of ti	0.60364	1.74206	9.343

Figure 24.2 Queuing Simulation

	A	B	C	D	E	F	G	H	I	J	K	L	M
1	EVENT#	EVENT TYPE	INTER-ARRIVAL TIME	SERVICE TIME	TM	SS	WL	AT	DT	#IN SYS	TIME BETWEEN EVENTS	Fraction of time busy	Average Number in system
2	0				0	0		0	9999	0		0.660931	1.766607
3	1	arrival	13.048364	9.3099643	0	1	0	13.04836	9.309964	1	9.309964312		
4	2	departure	6.5035581	5.4380628	9.309964	0	0	13.04836	9999	0	3.738399374		
5	3	arrival	15.610034	7.2353631	13.04836	1	0	28.6584	20.28373	1	7.23536307		
6	4	departure	1.8876646	7.6853583	20.28373	0	0	28.6584	9999	0	8.374670431		
7	5	arrival	7.3878094	6.346526	28.6584	1	0	36.04621	35.00492	1	6.346526008		
8	6	departure	1.9931364	0.3879305	35.00492	0	0	36.04621	9999	0	1.041283431		
9	7	arrival	12.626172	7.7931512	36.04621	1	0	48.67238	43.83936	1	7.793151226		
10	8	departure	5.2648076	3.0266887	43.83936	0	0	48.67238	9999	0	4.833020608		
11	9	arrival	17.673288	1.2835474	48.67238	1	0	66.34567	49.95593	1	1.283547418		
12	10	departure	12.495598	1.0014144	49.95593	0	0	66.34567	9999	0	16.3897401		
13	11	arrival	7.5488439	4.7645839	66.34567	1	0	73.89451	71.11025	1	4.76458391		
14	12	departure	3.9420927	1.4194422	71.11025	0	0	73.89451	9999	0	2.784259952		
31	29	departure	5.4932965	1.9080016	133.0251	1	0	133.4858	134.9331	1	0.460684731		
32	30	arrival	10.323678	2.1813543	133.4858	1	1	143.8094	134.9331	2	1.447316869		
33	31	departure	2.6472768	1.7987448	134.9331	1	0	143.8094	136.7318	1	1.798744811		
34	32	departure	0.9655007	4.6135214	136.7318	0	0	143.8094	9999	0	7.077616622		
35	33	arrival	6.7479613	2.5612785	143.8094	1	0	150.5574	146.3707	1	2.561278539		
36	34	departure	1.6392043	1.3052966	146.3707	0	0	150.5574	9999	0	4.186682772		
37	35	arrival	2.5290573	1.724248	150.5574	1	0	153.0864	152.2816	1	1.724248029		
38	36	departure	13.674124	3.2870862	152.2816	0	0	153.0864	9999	0	0.804809289		
39	37	arrival	0.9828946	9.3713517	153.0864	1	0	154.0693	162.4578	1	0.982894627		
40	38	arrival	1.6364608	4.4053739	154.0693	1	1	155.7058	162.4578	2	1.636460842		
41	39	arrival	18.261907	1.7575113	155.7058	1	2	173.9677	162.4578	3	6.751996237		
42	40	departure	18.671834	0.6611293	162.4578	1	1	173.9677	163.1189	2	0.661129287		
43	41	departure	0.8350733	1.1447845	163.1189	1	0	173.9677	164.2637	1	1.144784479		

Figure 24.2a

15 customers per hour. Interarrival times and service times are exponentially distributed. You want to estimate the fraction of the time the drive-through window is busy and the average number of customers present. Develop a spreadsheet that can be used to simulate 100 events (an event is an arrival or departure).

Solution The spreadsheet is given in Figure 24.2 (file Queue.wk4 or Queue.xls). All times are in minutes. We will advance the "clock time" of the simulation from 0 as the simulation progresses.

The columns contain the following information:

Event# is the number of the event (arrival or departure). We will simulate 100 events.

Event Type tells the type of event (arrival or departure) that has occurred.

Interarrival Time is the time between successive customer arrivals.

Service Time is the time required for service for successive customers.

TM is the current clock time.

SS is the status of server (0 = idle server and 1 = server busy).

WL is the number of customers in line.

AT is the time of the next arrival.

DT is the time of the next departure (if server is idle, we set this to 9999, or any number exceeding the maximum value of TM that can occur during the simulation.

#INSYS is the number of customers in the system.

To set up the simulation we proceed as follows:

Step 1 In cells A2:A102 we use the Data Fill option for Excel or Lotus to enter the number of the event (arrival or departure) (see Figure 24.2a).

Step 2 In C3 we enter the formula

```
Excel:  =RISKEXPON(6)

Lotus:  @RISKEXPON(6)
```

and copy it to the range C3:C102. This creates customer interarrival times that are independent and have a mean of 6 minutes.

Step 3 In D3 we enter the formula

```
Excel:  =RISKEXPON(4)

Lotus:  @RISKEXPON(4)
```

and copy it to the range D3:D102. This creates customer service times that are independent and have a mean of 4 minutes.

Step 4 In row 2 of the spreadsheet we initialize the simulation. Assuming that a customer has arrived at Time 0, we set TM = 0, SS = 0, AT = 0, and DT = 9999 (indicating that the server is currently idle).

Step 5 In row 3 we account for the first customer's arrival. The clock time is still 0 (so TM = 0). In F3 we enter the formula

```
Excel: =IF(J3>0,1,0)

Lotus: @IF (J3>0,1,0)
```

to determine whether the server is busy.

Step 6 In G3 we enter the formula

```
Excel: =MAX(J3-1,0)

Lotus: @MAX(J3-1,0)
```

to compute the number of people in line.

Step 7 In H3 we generate the next time a customer arrives by recopying the arrival time generated in cell C3 with the formula

```
Excel: =C3

Lotus: +C3
```

Step 8 In I3 we generate the departure time of the customer arriving at Time 0 by using his or her service time (determined in D3).

```
Excel: =D3

Lotus: +D3
```

Step 9 In J3 we update the number of people in the system to account for whether Event 1 was an arrival or departure. Actually, we know Event 1 is an arrival, but this step will help us with later events. We reason as follows: Event 1 will be an arrival if H2 < I2; otherwise Event 1 will be a departure.

Thus, entering in J3 the formula

```
Excel: =IF(H2<I2,J2+1,J2-1)

Lotus: @IF(H2<I2,J2+1,J2-1)
```

will correctly update the number of customers present.

Step 10 In row 4 we may now determine whether the second event is an arrival or departure. The time of the next event will be the minimum of AT (in H3) and DT (in I3). Thus, in E4 we compute TM by the formula

```
Excel:  =MIN(H3,I3)
Lotus:  @MIN(H3,I3)
```

Since $H3 > I3$, the event is a departure.

Step 11 In F4 we compute the status of the server with the formula

```
Excel:  =IF(J4>0,1,0)
Lotus:  @IF(J4>0,1,0)
```

Since the server is not busy, we obtain a 0 here.

Step 12 In G4 we compute the number of people in line by the formula

```
Excel:  =MAX(J4-1,0)
Lotus:  @MAX(J4-1,0)
```

This yields 0, because no people are in line.

Step 13 In H4 we compute the time the next arrival occurs. If the current (row 4) event is an arrival, then we find the time of the next arrival by using the interarrival time generated in C4 and adding it to the current clock time. If the current event is a departure, then the time of the next arrival remains unchanged. This logic is carried out by entering in cell H4 the formula

```
Excel:  =IF(H3<I3,H3+C4,H3)
Lotus:  @IF(H3<I3,H3+C4,H3)
```

Step 14 In I4 we determine the time the next service completion occurs. This requires that we consider four cases:

Case 1: If the current event is a departure ($H3 \geq I3$) and more than one person was present before the current event occurs, the next departure time will be the current clock time plus a new service time generated with the service time in D4 (this yields a departure time of $E4 + D4$).

Case 2: If the current event is a departure ($H3 \geq I3$) and one person was present before the current event occurs ($J3 = 1$), set the next departure time to 9999 (this indicates that the server is now empty, so a departure cannot occur until another arrival occurs).

Case 3: If the current event is an arrival ($H3 < I3$) and the server was empty before the current event (indicated by $I3 = 9999$), we determine the new departure time by adding the service time generated in D4 to the current clock time. This again yields a departure time of $E4 + D4$.

Case 4: In all other situations the time of the next departure remains unchanged (it is still I3).

Entering the following statement into I4 incorporates the logic of the four cases:

```
Excel: =IF(AND((H3>=I3),(J3>1)),E4+D4,
IF(AND((H3>=I3),(J3=1)),9999,IF((AND((H3<I3),
I3=9999)),E4+D4,I3)))

Lotus: @IF(H3>=I3#AND#J3>1,E4+D4,
@IF(H3>=I3#AND#J3=1,9999,
@IF(H3<I3#AND#I3=9999,E4+D4,I3)))
```

Step 15 In J4 we enter the formula

```
Excel: =IF(H3<I3,J3+1,J3-1)

Lotus: @IF(H3<I3,J3+1,J3-1)
```

to update the number of people present according to whether the current event is an arrival (H3 < I3) or a departure (H3 ≥ I3).

Step 16 Copying the formulas from B4:J4 to B4:J102 completes the simulation of 100 events.

Step 17 To indicate the type of each event (arrival or departure) we enter into cell B3 the formula

```
Excel: =IF(H2<I2,"arrival","departure")

Lotus: @IF(H2<I2,"arrival","departure")
```

Copying this formula to the range B4:B102 ensures that the type of each event will be recorded.

To see how things go, let's manually follow what happens in events 1–4. Event 1 is, of course, an arrival at 0. In C3 we generate the next customer's arrival to occur at 13.05, and in D3 we generate the first customer's service time to equal 9.31. Thus the next event (Event 2) is a departure. This brings us down to 0 people present. The next event (Event 3) must be, of course, an arrival. This will occur at time 13.05. To see what happens next we generate the instant of the next arrival (13.05 + 15.61 = 28.66) using the interarrival time created in C5. We generate the next service completion time (13.05 + 7.24 = 20.28) by using the service time in D5. Since 20.28 < 28.66, the next event (Event 4) is a departure.

24.1 Estimating the Operating Characteristics of a Queuing System

Two quantities of interest in an queuing system are

Server Utilization: The fraction of time a server is busy.

Expected Number in System: On the average, the expected number of customers present.

Figure 24.2b

Queuing Simulation					
	Figure 24.2				
	Cell	Name	Minimum	Mean	Maximum
	L2	/Fraction of time bus	0.415183	0.652602	0.9686575
	M2	/Fraction of time bus	0.60364	1.742062	9.343929

Example 24.1 continued

To estimate the fraction of the time that the drive-through window is busy we begin by computing the time between events. In K3 we enter

```
Excel: =E4-E3

Lotus: +E4-E3
```

This is the amount of time elapsing between Events 1 and 2. In L2 we can now compute the fraction of the time the server is busy with the formula

```
Excel: =SUMPRODUCT(K3:K101,F3:F101)/E102

Lotus: @SUMPRODUCT(K3..K101,F3..F101)/E102
```

`SUMPRODUCT(K3:K101,F3:F101)` gives the total time the server is busy, since it only counts when SS = 1.

To estimate the average number of customers present, we note that

Average Number of Customers Present
= 0 * (Fraction of Time 0 Customers are Present)
+ 1 * (Fraction of Time 1 Customer is Present)
+ 2 * (Fraction of Time 2 Customers are Present) + . . .

To compute this quantity enter into cell M2 the formula

```
Excel: =SUMPRODUCT(K3:K101,J3:J101)/E102

Lotus: @SUMPRODUCT(K3..K101,J3..J101)/E102
```

After selecting the cell range L2:M2 as our output range, we estimate that the teller will be busy 65.2% of the time and an average of 1.74 customers will be present (see Figure 24.2b). Using standard queuing calculations (see Chapter 22 of Winston 1994), it can be shown that in reality the teller will be busy 2/3 of the time and an average of 2 customers will be present.

Problems

Group A

24.1 Redo Example 24.1 with the following change: customers arrive every four minutes.

24.2 Redo Example 24.1 with the following change: it always takes two minutes to serve a customer.

24.3 Redo Example 24.1 with the following change: there is room for only two cars (including the car in service) at the drive-through window. Any car that arrives when two cars are present leaves without being served.

Group B

24.4 A doctor's office schedules patients at 15-minute intervals beginning at 9:00 and ending at 4:00. Patients are equally likely to arrive at any time within five minutes of their appointment. The number of minutes the doctor spends with a patient is governed by the distribution in Table 24.1.

Table 24.1

Time in Minutes	Probability
10	.60
20	.20
30	.20

a Estimate the probability that the doctor will be able to leave by 5:00.

b On the average, how many patients are present in the office?

References

Law, A., and Kelton, W. 1991. *Simulation Modeling and Analysis*. New York: McGraw-Hill.

Pegden, C., Shannon, R., and Sadowski, R. 1995. *Introduction to Simulation Using SIMAAN*. New York: McGraw-Hill.

Pritsker, A., Sigal, E., and Hammesfahr, J. 1994. *SLAM II: Network Models for Decision Support*. San Francisco: Scientific Press.

Schriber, T. 1991. *Simulation Using GPSS/H*. New York: John Wiley.

INDEX